HYPER-STRUCTURED MOLECULES II
CHEMISTRY, PHYSICS AND APPLICATIONS

Edited by

Hiroyuki Sasabe
CREST, JST and RIKEN
Saitama, Japan

Gordon and Breach Science Publishers

Australia • Canada • France • Germany • India • Japan
Luxembourg • Malaysia • The Netherlands • Russia • Singapore • Switzerland

Amsteldijk 166
1st Floor
1079 LH Amsterdam
The Netherlands

British Library Cataloguing in Publication Data

Hyper-structured molecules II: chemistry, physics and applications
  1. Macromolecules
  I. Sasabe, Hiroyuki
  547.7

ISBN 90-5699-215-5

CONTENTS

CONTENTS

PREFACE

Under the strategic research promotion program by the Japan Science and Technology Corporation, the Core Research for Evolutional Science and Technology (CREST) program was launched in March, 1996. The Hyper-Structured Molecules and their Application to Organic Quantum Devices group was approved as a member of the CREST program for five years. The aims of this group are to design molecules such as dendritic polymers and topologically controlled molecules and to handle these molecules by 'molecular tweezers' such as photon STM. Hyper-structured molecules are topologically well-defined molecules in two and/or three dimensions, and expected to show novel quantum effects in the molecules themselves and in molecular sequences, e.g., light emitting molecules, chemical amplification, and molecular magnets. The 1st International Forum on Hyper-Structured Molecules was held in Kusatsu, Japan, on 4–6 November 1996, and the 2nd in Sapporo, Japan, on 30 May–1 June 1997, to discuss the possibility and future of these new molecules for real application to organic quantum devices. The forum will be held every year until the end of the program.

This monograph on *Hyper-Structured Molecules* is the collected papers of the second forum and covers molecular designs of dendrimers, oligomers, hyperbranched polymers and/or high spin systems, molecular organizations and nanostructures, mesoscopic pattern formation, and scanning probe microscopy for characterization and molecular handling.

I would like to express my gratitude to all the contributors to the forum who activated the new project as a most promising research field. The financial support by the Japan Science and Technology Corporation is also highly appreciated. Finally I would like to thank Gordon and Breach Science Publishers for their continuous encouragement for this monograph.

Hiroyuki Sasabe
Project Leader

LIST OF CORRESPONDING AUTHORS

Kunio Awaga
Department of Basic Sciences, The University of Tokyo, 3-8-1 Komaba,
Meguro-ku, Tokyo 153-8902, Japan
TEL: +81-3-5454-6750   FAX: +81-3-5454-6768
E-mail: awaga@awg.c.u-tokyo.ac.jp

Donald Bethell
Department of Chemistry, University of Liverpool, PO Box 147,
Liverpool L69 3BX, UK
TEL: +44-151-794-3508   FAX: +44-151-794-3508
E-mail: bethell@liverpool.ac.uk

Guido Eilers
Nanoelectronics Laboratory, Graduate School of Engineering, Hokkaido
University, Kita 13, Nishi 8, Kita-ku, Sapporo 060-8628, Japan
TEL: +81-11-706-6539   FAX: +81-11-706-7803
E-mail: guido@nano.eng.hokudai.ac.jp

Eiji Hatta
Nanoelectronics Laboratory, Graduate School of Engineering, Hokkaido
University, Kita 13, Nishi 8, Kita-ku, Sapporo 060-8628, Japan
TEL: +81-11-706-6539   FAX: +81-11-706-7803
E-mail: hatta@nano.eng.hokudai.ac.jp

Takashi Isoshima
RIKEN (The Institute of Physical and Chemical Research), 2-1 Hirosawa,
Wako, Saitama 351-0198, Japan
TEL: +81-48-462-1111 ext. 6325   FAX: +81-48-462-4695
E-mail: isoshima@postman.riken.go.jp

Ruggero Micheletto
Frontier Research Program, RIKEN (The Institute of Physical and
Chemical Research), 2-1 Hirosawa, Wako, Saitama 351-0198, Japan
TEL: +81-48-462-1111 ext. 6336   FAX: +81-48-462-4695
E-mail: ruggero@postman.riken.go.jp

Koichi Mukasa
Nanoelectronics Laboratory, Graduate School of Engineering, Hokkaido
University, Kita 13, Nishi 8, Kita-ku, Sapporo 060-8628, Japan
TEL: +81-11-706-6539   FAX: +81-11-706-7803
E-mail: mukasa@nano.eng.hokudai.ac.jp

Shigeru Murata
Department of Basic Sciences, The University of Tokyo, 3-8-1 Komaba,
Meguro-ku, Tokyo 153-8902, Japan
TEL: +81-3-5454-6596   FAX: +81-3-3485-2904
E-mail: cmura@komaba.ecc.u-tokyo.ac.jp

Kohji Nakamura
Department of Physics and Astronomy, Northwestern University,
2145 Sheridan Road, Tech F263, Evanston, Illinois 60208-3112, USA
TEL: +1-847-491-8640   FAX: +1-847-491-5082
E-mail: kohji@pluto.phys.nwu.edu

Masahisa Osawa
Organometallic Chemistry Laboratory, RIKEN (The Institute of Physical
and Chemical Research), 2-1 Hirosawa, Wako,
Saitama 351-0198, Japan
TEL: +81-48-462-1111 ext. 3515   FAX: +81-48-462-4665
E-mail: osawa@postman.riken.go.jp

Hiroki Oshio
Department of Chemistry, Graduate School of Science,
Tohoku University, Aoba, Aramaki, Aoba-ku,
Sendai 980-8578, Japan
TEL: +81-22-217-6545   FAX: +81-22-217-6548
E-mail: oshio@agnus.chem.tohoku.ac.jp

Makoto Sawamura
Nanoelectronics Laboratory, Graduate School of Engineering, Hokkaido
University, Kita 13, Nishi 8, Kita-ku, Sapporo 060-8628, Japan
TEL: +81-11-706-6539   FAX: +81-11-706-7803
E-mail: sawamura@nano.eng.hokudai.ac.jp

Masatsugu Shimomura
Research Institute for Electronic Science, Hokkaido University,
Kita 12, Nishi 6, Kita-ku, Sapporo 060-0812, Japan
TEL: +81-11-706-2997   FAX: +81-11-706-4974
E-mail: shimo@imdes.hokudai.ac.jp

Tadashi Sugawara
Department of Basic Sciences, The University of Tokyo, 3-8-1 Komaba,
Meguro-ku, Tokyo 153-8902, Japan
TEL: +81-3-5454-6742   FAX: +81-3-3481-5934
E-mail: suga@pentacle.c.u-tokyo.ac.jp

Tatsuo Wada
Supramolecular Science Laboratory, RIKEN (The Institute of Physical and
Chemical Research), 2-1 Hirosawa, Wako, Saitama 351-0198, Japan
TEL: +81-48-467-9378   FAX: +81-48-462-4647
E-mail: tatsuow@postman.riken.go.jp

Michael R. Wasielewski
Chemistry Division, Argonne National Laboratory, Argonne,
Illinois 60439-4831, USA
TEL: +1-630-252-3538   FAX: +1-630-252-9289
E-mail: wasielewski@anlchm.chm.anl.gov

Karen L. Wooley
Department of Chemistry, Campus Box 1134, Washington University,
One Brookings Drive, St Louis, Missouri 63130-4899, USA
TEL: +1-314-935-7136   FAX: +1-314-935-4481
E-mail: klwooley@artsci.wustl.edu

Shiyoshi Yokoyama
Communications Research Laboratory, Kansai Advanced Research Center,
588-2 Iwaoka, Nishi-ku, Kobe 651-2401, Japan
TEL: +81-78-969-2254   FAX: +81-78-969-2259
E-mail: syoko@crl.go.jp

1. HYPER-STRUCTURED MOLECULES FOR PHOTONIC APPLICATIONS

TATSUO WADA[a,b,*], YADONG ZHANG[a] and HIROYUKI SASABE[a,c]

[a]Core Research for Evolutional Science and Technology (CREST), JST
[b]Supramolecular Science Laboratory, RIKEN (The Institute of Physical and Chemical Research), 2-1 Hirosawa, Wako, Saitama 351-0198, Japan
[c]Department of Photonic Materials Science, Chitose Institute of Science and Technology, 758-65 Bibi, Chitose, Hokkaido 066-8655, Japan

ABSTRACT

In order to fill up the gap between molecular design and architecture of three-dimensional structures in molecular electronic devices, we propose the novel idea of hyper-structured molecules (HSMs). We have developed carbazole dendrimers and oligomers as HSMs, and selected photorefractive effects as a primary target function based on their molecular functionalities. Oligomers developed in our laboratory exhibit photocarrier generation and transporting, electro-optic, film-forming and poling properties. These multifunctional properties allow us to demonstrate optical image processing using optical phase conjugation.

FROM NANOSTRUCTURES TO MEDs

A great deal of interest surrounds the creation of materials with characteristic dimensions of the order of the wavelength of an electron. The fabrication of nanometer scale structures has typically been followed by two different approaches: a "downsizing" of macroscopic materials and a "building-up" of molecular blocks. The former approach, especially in inorganics, is the basis for well-developed photolithographic patterning and etching techniques. The latter, especially in the case of organics, is the basis for molecular and/or atom manipulation as illustrated in Figure 1. Successful progress in microfabrication techniques have allowed us to produce integrated electronic circuits for high performance central computing units and high density memories in the field of semiconductor electronic devices. However the limitation of photolithography has been raised in the Si-based device fabrication. Following this problem, there have been several attempts

---

* Tel.: +81-48-467-9378, Fax: +81-48-462-4647, E-mail: tatsuow@postman.riken.go.jp

● INORGANICS:

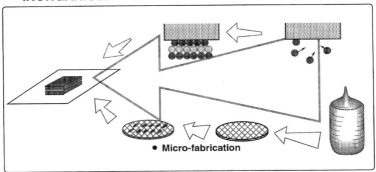

● ORGANICS:

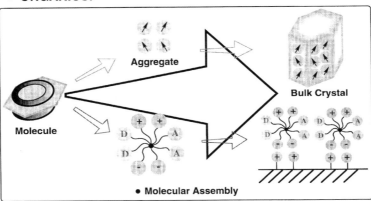

Figure 1 Approach to nanostructures.

beyond Si-electronic devices: new processing techniques for atom manipulation and new device concepts including non-Si materials. Instead of nanoscale silicon devices, organic molecules may be used as a device element based on molecular level switching. The concept of "Molecular Electronic Devices (MEDs)" was proposed by Dr. Carter in early 1980.[1] These MEDs require proper molecular design and assembly on the circuitry board (solid substrate). Various new techniques to handle functional molecules have been developed including Langmuir-Blodgett (LB) techniques. For example, in order to actualize a molecular rectifier,[2] the LB technique was applied to align molecules with a donor and acceptor moiety connected by a non-conjugated bridge, but there is still a wide gap to perfect a molecular level device. After the death of Dr. Carter in 1987, the enthusiasm toward

MEDs declined steadily because of the difficulties in handling and/or addressing the molecular components. Recently there has been remarkable progress in scanning tunneling microscopy (STM) techniques. The demonstration of the atom-manipulation using STM by Dr. Eigler (IBM Almaden) opens up the possibility of communication and manipulation of single molecules.[3]

STRUCTURE AND FUNCTIONS

In the case of biological molecular systems, higher order structures of protein provide the special coordination of active sites and consequently efficient energy and/or electron transfer processes are achieved. These biological functional structures are based on the specific interaction of functional groups in amino acid residual groups in proteins. There have been a lot of synthetic efforts to mimic the biological systems. Even though we can control the primary structures and tacticities of polypeptides by synthetic chemistry approach, there is a large gap between molecular design and the three-dimensional (3-D) structure as illustrated in Figure 2.

The molecular building blocks can order themselves and condense to form higher-order structures in a specific manner. Strength and anisotropy of interactions between subunits, and symmetry of packing are major organizing parameters in inorganics. On the other hand, organics have a variety of characteristics such as symmetry of molecule, molecular polarizability, hydrophilic and hydrophobic properties, and polarity of substituents. These unique features are based not only on the nature of the molecules but also "molecular subunits" can be used as organizing parameters in organics. While the optoelectronic properties of the condensed state can be derived from those of the isolated molecules, the organization of molecular building blocks also affects on those of the condensed state. Namely we have to incorporate not only the organizing parameters but also optoelectronic functions into the molecular building blocks in order to construct a functional 3-D structure. Taking these ideas into account, we have utilized not a point-like molecule but oligomers as a building block which offer more flexibility when building the functional 3-D structure.

HYPER-STRUCTURED MOLECULES

To fill up the gap between molecular design and architecture of 3-D structures, we proposed the idea of hyper-structured molecules (HSMs) in which molecules themselves have well-defined and specific structures.[4] In other words, the molecular design of HSMs is directly related to the

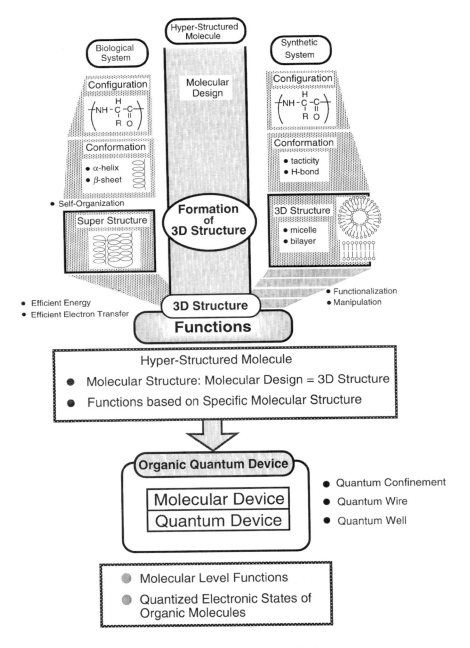

Figure 2 Approach to HSMs and organic quantum devices.

realization of the specific 3-D structure without molecular assembly as illustrated in Figure 2. Furthermore, functions in the molecular level can be expected in HSMs based on their specific 3-D structure. A molecule with a "topologically well-defined structure" such as a dendrimer is one of the candidates for HSMs. HSMs have more than two molecular units which must be coordinated or organized for particular purposes to create a specific structure using various bond arrangements including covalent bond.

Dendritic polymers or starburst polymers have attracted a lot of interests based on their peculiar structure. Using the convergent method, we can prepare asymmetric dendrimers which allow us to control the charge distribution of molecules or local reactivity such as hydrophilic, hydrophobic, electron-donative and electron-acceptive properties. There is a special article on the potential research activity of dendritic polymers and dendrimers entitled "Translating beauty into function" in C&E News.[5] Several research groups have developed functional dendrimers: triphenyl amine dendrimers for stable electroluminescent materials;[6] metallodendrimers;[7] liquid crystals.[8] By the stepwise precise syntheses of dendrons, the number of terminal groups and molecular weight can be controlled as shown in Figure 3. The molecular size and 3-D shape can also change depending on the generation (I, II, III in Figure 3) of the dendrons. The diameter of the 5-generation-dendrimer covers around 10 nm,[9] and the shape of this dendrimer changes

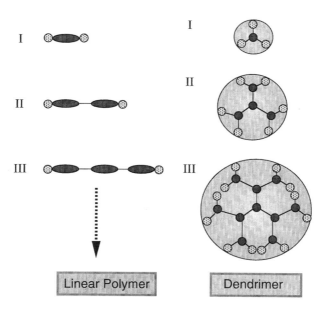

Figure 3 Generation of dendrimers and linear oligomers.

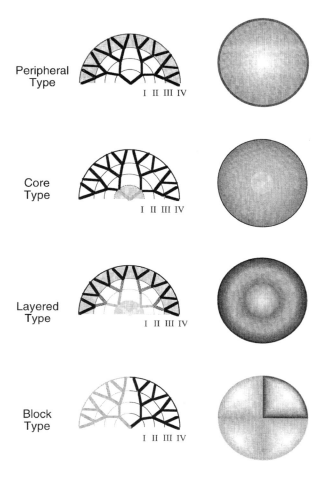

Figure 4   Different types of functional dendrimers.

from oval to sphere (elucidated by carbon-13 spin-lattice relaxation measurement[10]).

One of the most interesting and important features in the dendrimer structure is that the number of nodes increases depending on the generations as shown in Figure 3. In contrast to linear chain oligomers, a large number of terminal positions are possible. We can introduce electroactive groups into the terminal positions of dendrimers, then we can modify the total features of dendrimers chemically and physically. While there is structural freedom in linear chain oligomers, the molecular shape of dendrimers can be controlled by the structure of the spacer units and the generation. Using rigid spacer groups, we can obtain stretched linear oligomers, however, these rod-like molecules show poor processability. On the other hand, it was reported that higher generation dendrimers are quite soluble in common organic solvents. Besides the feasibility of chemical modification, interesting physical properties can be expected in the dendrimer structures.

Figure 4 illustrates examples of the functional dendrimers. That is, we can introduce a variety of functional groups in a specific position. There are four possible types of functional dendrimers depending upon where the most functional groups are located as shown in Figure 4: (1) peripheral type: various electroactive and/or reactive groups were introduced on the peripheral part; (2) core type: photoactive groups are sterically surrounded by dendrons; (3) layered type: different types of electroactive groups such as an electron acceptor and donor are introduced on the node positions (each generation); (4) block type: each block of dendron has different functional groups. In these dendrimers we can obtain asymmetric charge distribution, directional charge transfer and/or anisotropy of reactivity.

SUPRAMOLECULES, SUPERMOLECULES AND HSMs

In support of the creation of superior functional molecular systems, various supramolecules have been proposed.[11] Supramolecules may have advantages beyond simple point-like molecules or polymers. There are a lot of similarities between supramolecules and HSMs, however, strictly speaking, dendrimers composed by only covalent bonds are not supramolecules. They might better be called "superatoms", and would form "supermolecules" from the asymmetric dendrimers by non-covalent interactions. The dendrimers in which subunits or dendrons are constructed by non-covalent bonds can be classified as supramolecules.

The problem of how to handle dendrimer molecules should be addressed for the preparation of supermolecules. Along with the recent development of scanning probe techniques, the handling and detection of single molecules have become reality. STM techniques can be applied to manipulate and communicate HSMs. In comparison to point-like molecules, HSMs have several advantages in terms of their larger molecular size and well-defined structure for STM techniques. Laser trapping is one method of handling macromolecules micrometers in size. If near field optics is applied to single molecules, then the optical evanescent field interacts with the molecules resulting in a dielectric force. This force moves the molecule. Based on this idea, HSMs can be translocated along the optical channel waveguide formed on a solid substrate (e.g., glass substrate). Near field optics with high resolution also allow us to access the HSMs and carry out single molecular spectroscopy.

Electro- and/or photoactive sites of dendrimers are organized spatially in a specific position which can provide an efficient electron or energy transfer. Individual HSMs can also behave as a quantum dot (quantum confinement) if designed in the form of a layered type or core type structure (shown in Figure 4). These molecules have a strong exciton-exciton interaction inside the molecule and/or transfer electrons generated by photons in the outer shell to the core, which behave as a light-harvesting system or a reaction center in photosynthesis. In addition, one-dimensional arrays of block type HSMs stabilized on the solid substrate may transfer electrons injected from the "acceptor" portion of the molecule,[4] which is a model of the quantum wire. Input and output of electrons are easily performed by two STM tips. Another example is the formation of a quantum well structure when the different HSMs are arranged periodically in one dimension.

MULTIFUNCTIONAL CARBAZOLE OLIGOMERS FOR MONOLITHIC PHOTOREFRACTIVE MATERIALS

Figure 5 illustrates the current status toward HSMs in our research project. We have selected multifunctional oligomers as a concrete example of HSMs and photorefractive effects as a primary target of functions based on their molecular multifunctionalities. Since the diffraction of light caused by the refractive index gratings is a bulk response, photorefractive effects require multifunctional properties, *i.e.*, both photoconductivity and electro-optic responses in the materials. This is one of the interesting and challenging targets for the demonstration of multifunctional performance of HSMs, especially functional dendrimers and oligomers.

Photorefractive materials are the key materials in future optical systems such as a real-time holography for dynamic optical storage.[12] Ferroelectric crystals, *e.g.*, $BaTiO_3$ and $KNbO_3$, are typical photorefractive materials and have a unique set of principal axes. Since they must be used in bulk form, very careful crystal growth and processing are usually required. Organic materials, on the other hand, can form thin films easily. For this reason there is excitement with the recent development of new organic thin films which exhibit photorefractive properties after poling.[13] In order to exhibit photorefractive effects, materials must meet several requirements: second-order optical nonlinearity to provide Pockels effect and photoconductivity to create a space charge. The first demonstration of photorefractive effects in organic materials was in charge-transfer complex crystals: noncentro-symmetric organic crystals slightly doped with an electron acceptor.[14] After this, several research groups reported photoinduced refractive index changes in poled photoconductive polymers.[15] These photorefractive polymers generally consist of three components: the first is a nonlinear optically active

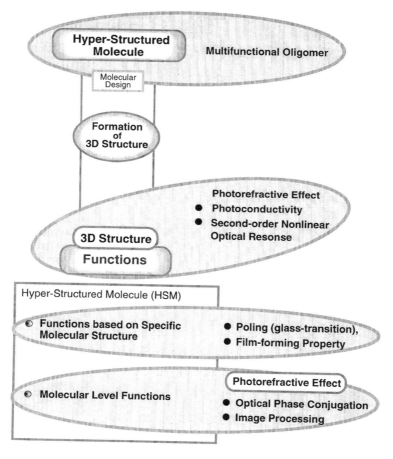

Figure 5 Current status of HSMs project.

chromophore to provide an electro-optic response, the second is a carrier (usually hole) transporting molecule and the third is a photosensitizer to enhance a carrier generation, as summarized in Figure 6. The last two components provide photoconductive properties. The requirements for photorefractivity in organic materials are usually achieved by using multi-components, each of which fulfills a single requirement as illustrated in Figure 7. In the case of multifunctional materials, the polymer backbone offers film-forming properties, together with electro-optic response and photoconductive properties in the side chains. Recently remarkable progress has been made in improving the photorefractive performance in multi-component photorefractive polymers by addition of a plasticizer which

TATSUO WADA *et al.*

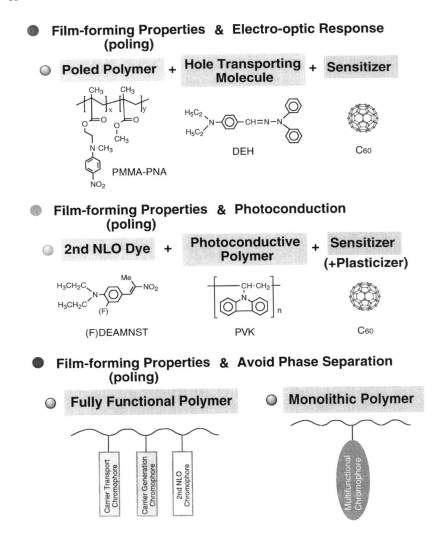

● **Film-forming Properties & Electro-optic Response (poling)**

○ **Poled Polymer** + **Hole Transporting Molecule** + **Sensitizer**

PMMA-PNA   DEH   C₆₀

● **Film-forming Properties & Photoconduction (poling)**

○ **2nd NLO Dye** + **Photoconductive Polymer** + **Sensitizer (+Plasticizer)**

(F)DEAMNST   PVK   C₆₀

● **Film-forming Properties & Avoid Phase Separation (poling)**

○ **Fully Functional Polymer**   ○ **Monolithic Polymer**

Carrier Transport Chromophore | Carrier Generation Chromophore | 2nd NLO Chromophore   Multifunctional Chromophore

Figure 6 Various organic photorefractive materials.

reduces the glass transition temperature (T_g).[16] Although large photo-induced refractive index changes have been obtained in low-T_g multi-component polymers, there is a limitation on maximum concentration of chromophores because of phase separation and crystallization. In order to overcome this problem, we have developed multifunctional chromophores which enable us to obtain "monolithic" photorefractive materials.[17]

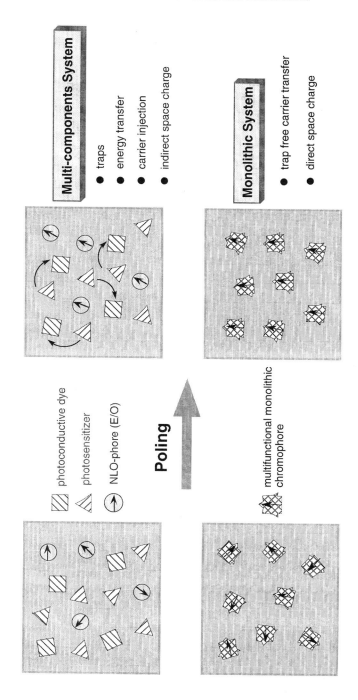

Figure 7 Multicomponent and monolithic photorefractive materials.

The single component of multifunctional chromophores fulfills all of the requirements for photorefractivity as shown in Figure 7.

We selected an acceptor-substituted carbazole as the building block for multifunctional chromophores.[17] Carbazole derivatives are well-studied hole-transporting molecules and their photocarrier generation efficiency can be sensitized by the formation of a charge-transfer complex. The carbazole molecule has an isoelectronic structure of diphenylamine. Therefore, the introduction of acceptor groups in 3 and/or 6-position induces intramolecular charge transfer. The poling behavior of absorption spectra was examined experimentally and theoretically for one- and two-dimensional charge-transfer systems based on monosubstituted and disubstituted carbazole derivatives.[18]

The results of photoconductive and electro-optic experiments confirm the dual functionality of acceptor-substituted carbazoles which fulfill the requirements of photorefraction. It should be noted that an alternative to our approach is through fully functionalized polymers in which multi-component moieties are incorporated into polymer side chains or main chains[19] (see Figure 6). These fully functionalized polymers including monolithic polymers have been developed as a macroscopic material and successfully suppressed the phase separation and increased the density of functional groups. However, they have distributions of molecular weight, number and size of free volume, and molecular structures. Moreover, there is a multi-step carrier hopping process among multicomponent moieties, which reduces the carrier mobility due to various traps.

On the other hand, the oligomeric molecules have a perfectly defined structure. Recently we synthesized various kinds of carbazole oligomers[20] such as cyclic oligomers, head-to-tail carbazole dimers, conjugated trimers and dendrimers besides main-chain polymers and hyper-branched polymers. The cyclic oligomer has alternating units of acceptor-introduced and electron donative carbazole moieties connected through 3,6-linkages. It should be noted that acceptor-introduced carbazoles are very promising for second-order nonlinear optical chromophores. We also synthesized carbazole dendrimers and conjugated trimers as multifunctional chromophores.[17] Since these oligomers have a highly branched structure with long spacer groups, they have film-forming properties and show glass-rubber transition. The T_g changes depending on the length of spacer groups and chromophore structures. Typical differential scanning calorimetry (DSC) traces of carbazole dendrimers with different acceptor groups are shown in Figure 8. Amorphous solid films can be formed by spin coating from a solution, and poled at an elevated temperature above T_g. The electric field-induced alignment of the chromophores were confirmed by second-harmonic generation (SHG) measurement at room temperature. With no electric field applied, the second-harmonic (SH) intensity was zero, as a result of the

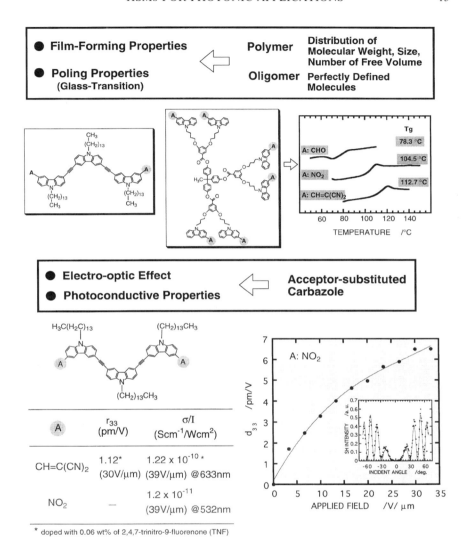

Figure 8 Multifunctionality in carbazole oligomers: glass-transition temperature (T_g), electro-optic coefficient (r_{33}), photoconductive sensitivity (σ/I), and second order nonlinear coefficient (d_{33}).

centrosymmetric random arrangement of the chromophores. After switching on the electric field, noncentrosymmetric alignment of the chromophores was achieved, reaching a stable plateau value within a few seconds.[21] The second order nonlinear coefficient (d_{33}) at a fundamental wavelength of

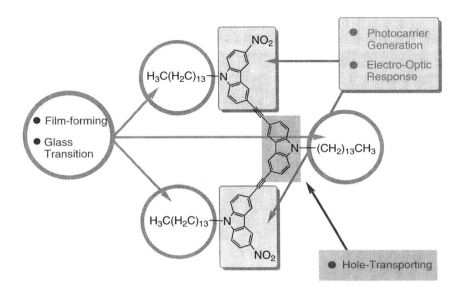

Figure 9 Molecular and structural design of multifunctional carbazole trimers.

1064 nm increased monotonously with an applied electric field as shown in Figure 8. Electro-optic coefficient (r_{33}) and photoconductive sensitivity (σ/I) are also summarized in Figure 8.

Figure 9 illustrates the molecular and structural design of the carbazole trimer for multifunctional properties. In this trimer, carbazole rings are linked to each other by an acetylene bond and peripheral carbazoles are substituted with a nitro group. Acceptor-substituted carbazole moieties provide a photocarrier generation and an electro-optic response. Central carbazole moiety has hole-transporting properties, and forms a charge-transfer complex with acceptor molecules such as 2,4,7-trinitro-9-fluorenone (TNF).[22] A conjugated structure prefers to exhibit a relatively high photo-carrier mobility.[23] In order to obtain a conjugated carbazole trimer with low T_g and film-forming properties, a long aliphatic group was introduced on the 9-position of each carbazole ring. Due to the attachment of the long aliphatic chain the carbazole trimer is amorphous and has T_g of 20 °C.

Photorefractive properties are characterized by two-beam coupling and four-wave mixing techniques. In the two-beam coupling measurement an asymmetric energy exchange between the two beams was observed when the electric field was applied. This also proved that a true photorefractive effect is present.[13] From this energy exchange experiment, the two-beam coupling gain coefficient (Γ) could be obtained from two-beam coupling ratios. Γ increased monotonously with an applied electric field. At an applied electric field of 33 V/μm, the photorefractive gain of 35.0 cm^{-1} was obtained. This gain is larger than the absorption coefficient ($\alpha = 8.2$ cm^{-1}) of this material, leading to a net two-beam coupling gain coefficient ($\Gamma - \alpha$) of 26.8 cm^{-1}. This value is superior to that of BaTiO$_3$ ($\Gamma = 2.2$ cm^{-1}) although conditions are not identical. It is well known that the majority of applications for photorefractive materials requires net gain. An applied electric field plays an important role in enhancing the photorefractive efficiencies because of the better alignment of the second-order nonlinear optical chromophores and higher photoconductivity at the higher electric field. If the external electric field is not applied during writing in a two-beam coupling experiment, no detectable refractive index gratings are observed due to the centrosymmetric random distribution of the second-order nonlinear chromophores for the low-T_g material.

A four-wave mixing technique was used to determine the steady-state diffraction efficiency. In Figure 10, the two s-polarized beams were used as the writing beams. The grating was read out with a weak p-polarized beam counter-propagated to one of the writing beams. A polarizing beam splitter was used to separate diffracted light from the counter propagating writing beam. The diffracted (phase conjugated) light was detected by a photodiode. A strong electric field dependence of the diffraction efficiency was also observed. At an applied electric field of 33 V/μm, a diffraction efficiency of 13% was obtained and a photoinduced refractive index change of 3.6×10^{-4} was calculated. These efficient photorefractive effects allow us to demonstrate optical image processing. Figure 10 also shows the phase conjugation via four-wave mixing and a demonstration of image processing.[24] The original image (Figure 10(a)) was distorted by the insertion of a phase aberrator. The wave front of distorted image (Figure 10(b)) was reconstructed via optical phase conjugation as shown in Figure 10(c). We have developed "monolithic photorefractive materials" using multifunctional chromophores and demonstrated their capability to perform image reconstruction using low power laser sources.

CONCLUSION

We have proposed the novel idea of hyper-structured molecules (HSMs) in which the molecular design is directly related to the realization of the specific three-dimensional structure. As a typical example of HSMs, we have developed carbazole dendrimers and trimers which exhibit multifunctional properties. These novel molecular systems allow us to obtain monolithic photorefractive materials as a new class of photorefractive materials. We believe the approach using hyper-structured molecules opens new possibilities for photonic applications.

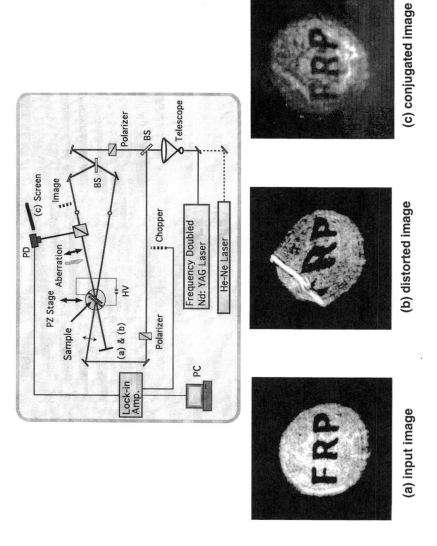

Figure 10 Optical phase conjugation using four-wave mixing: (a) original image; (b) distorted image; (c) conjugated image.

REFERENCES

1. F.L. Carter (1982). *Molecular Electronic Devices*, Marcel Dekker, New York, 5.
2. A. Aviram and M.A. Ratner (1974). *Chem. Phys. Lett.*, **29** 277.
3. D.M. Eigler and E.K. Schweizer (1990). *Nature*, **344**, 524.
4. H. Sasabe (1999). *Hyper Structured Molecules I-Chemistry, Physics and Applications*, H. Sasabe (Ed.), Gordon and Breach Sci. Pub., 1.
5. R. Dagani (1996). *Chem. & Eng. News*, June 3, 30.
6. Y. Shirota, Y. Kuwabara, H. Inada, T. Wakimoto, H. Nakada, Y. Yonemoto, S. Kawami and K. Imai (1994). *Appl. Phys. Lett.*, **65**, 807.
7. E.C. Constable (1997). *Chem. Commun.*, 1073.
8. V. Percec and M. Kawasumi (1992). In Polym. Prepr. (Am. Chem. Soc., Div. Polym. Chem.) **33**, 221.
9. D.A. Tomalia, A.M. Naylor and W.A. Goddard III (1990). *Angew. Chem. Int. Ed. Engl.*, **29**, 138.
10. A.M. Naylor, W.A. Goddard III, G.E. Kiefer and D.A. Tomalia (1989). *J. Am. Chem. Soc.*, **111**, 2339.
11. J.-M. Lehn (1995). *Supramolecular Chemistry* (VCH, Weinheim, New York, Basel, Cambridge, Tokyo.
12. D.G. Feitelson (1988). *Optical Computing*, MIT Press, Cambridge, 155.
13. W.E. Moerner and S.M. Silence (1994). *Chem. Rev.*, **94**, 127.
14. K. Sutter, J. Hulliger and P. Günter (1990). *Solid State Commun.*, **74**, 867.
15. S. Ducharme, J.C. Scott, R.J. Twieg and W.E. Moerner (1991). *Phys. Rev. Lett.*, **66**, 1846.
16. K. Meerholz, B.L. Volodin, Sandalphon, B. Kippelen and N. Peyghambarian (1994). *Nature*, **371**, 497.
17. T. Wada, L. Wang, Y. Zhang, M. Tian and H. Sasabe (1996). *Nonlinear Optics*, **15**, 103.
18. T. Isoshima, T. Wada, Y.-D. Zhang, E. Brouyère, J.-L. Brédas and H. Sasabe (1996). *J. Chem. Phys.*, **104**, 2467.
19. L. Yu, W. Chan, Z. Bao and X.F. Cao (1992). *Macromolecules*, **26**, 2216.
20. T. Wada, Y. Zhang and H. Sasabe (1997). *RIKEN Review*, **15**, 33.
21. Y. Zhang, L. Wang, T. Wada and H. Sasabe (1997). *Appl. Phys. Lett.*, **70**, 2949.
22. L. Wang, Y. Zhang, T. Wada and H. Sasabe (1996). *Appl. Phys. Lett.*, **69**, 728.
23. C. Beginn, J.V. Grazulevicius and P. Strohriegel (1994). *Macromol. Chem. Phys.*, **195**, 2353.
24. T. Wada, Y. Zhang, T. Aoyama and H. Sasabe (1997). *Proc. Japan Acad.*, **73**, Ser. B, 165.

2. THE DESIGN, SYNTHESIS, CHARACTERIZATION AND DERIVATIZATION OF FLUORINE-CONTAINING DENDRIMERS AND HYPERBRANCHED POLYMERS

THOMAS A. STRAW*, ANJA MUELLER,
ANDREI D. STEFANESCU, TOMASZ KOWALEWSKI
and KAREN L. WOOLEY**

*Department of Chemistry, Washington University, One Brookings Drive,
St Louis, Missouri 63130-4899, USA*

INTRODUCTION

Dendritic macromolecules have received considerable interest over the past decade, due to the unusual macroscopic properties that have been observed for the bulk materials (Tomalia, 1990; Newkome, 1994; Fréchet, 1994). The term dendritic means tree-like and dendritic macromolecules are highly-branched structures, containing at least one branching site at each monomeric repeat unit. The high degree of branching limits the extension of the molecule, requiring that the molecule occupy a smaller volume than for a corresponding linear polymer and that it adopts a globular shape. The branching and globular shape are believed to be the structural properties that prevent extensive chain entanglements to produce the unique properties (*e.g.*, low viscosities, high solubility, amorphous material) observed for dendritic macromolecules, in comparison to linear polymers (Hawker, 1997). To aid in the advancement of the field of dendritic macromolecules and to gain the ability to create specialized materials designed for superior performance in a variety of applications, thorough understanding of the dendrimer structure from the molecular level must be obtained.

In particular, we were interested in the determination of the size, shape, conformation and packing of dendrimers in the solid state. For example, although the low solution and melt viscosities implied a lack of extensive chain entanglements, direct measurement had not been made. By application of site-specific stable isotope labeling of the chain ends (^{13}C) and focal point (^{19}F) of benzyl ether dendrimers, the distances between the multiple ^{13}C-labeled chain ends and the single ^{19}F focal point (core) were evaluated

* Present Address: Courtaulds Corporate Technology, Coventry, UK.
** To whom correspondence should be addressed.

by rotational-echo double-resonance (REDOR) NMR (Wooley, 1997). The data for the third, fourth, and fifth generation dendrimers indicated that inward folding of the chain ends and interpenetration between dendrimers were occurring. The extent of interpenetration decreased with increasing generation number. Based upon these results, enhanced reduction of the cohesive forces between the dendrimer molecules was attempted to produce materials that would behave as non-interacting nanosized particles.

Fluorocarbons are well known to exhibit decreased intermolecular attractive forces, and increased hydrophobicity, lipophobicity, chemical resistance and thermal stability, in comparison to their hydrocarbon analogs (Chambers, 1973). Fluoropolymers possess these same properties and they find broad application as minimally-adhesive materials due to their low surface energies (Smart, 1994; Brady, 1990; Gangal, 1989). Therefore, fluorocarbon units were placed at the chain ends of dendrimers and the physical and mechanical properties were investigated. In order to accurately evaluate the effects of the fluorocarbon content and also the structural branching, the fluorocarbon-containing dendrimers were compared with non-fluorinated dendrimers and also fluorine-containing hyperbranched polymers (highly-branched polymers with degrees of branching less than 100% and typically *ca.* 50%).

RESULTS AND DISCUSSION

Dendrimers

Synthesis

Dendritic structures are constructed by a "bottom-up" approach, in which repetitive multi-step syntheses are used. The syntheses begin either at the core (divergent) or at the chain ends (convergent), with consecutive layers of branching monomeric repeat units added with each growth step (generation). The multi-step routes allow for purification at each generation to ensure that the highest degree of perfection and branching be accomplished.

Pentafluorophenyl groups were selected as the fluorocarbon units to be placed at the dendrimer chain ends, due to their compatibility with the chemistry used for the construction of benzyl ether dendrimers and the possibility for modification of the dendrimers once prepared, through reactions of the pentafluorophenyl moieties. The convergent growth approach allowed for the placement of the pentafluorophenyl groups at the chain ends in the first step of the synthesis (Scheme 1). Reaction of penta-fluorobenzyl bromide with 3,5-dihydroxylbenzyl alcohol in the presence of potassium carbonate and 18-crown-6 in tetrahydrofuran gave alkylation of

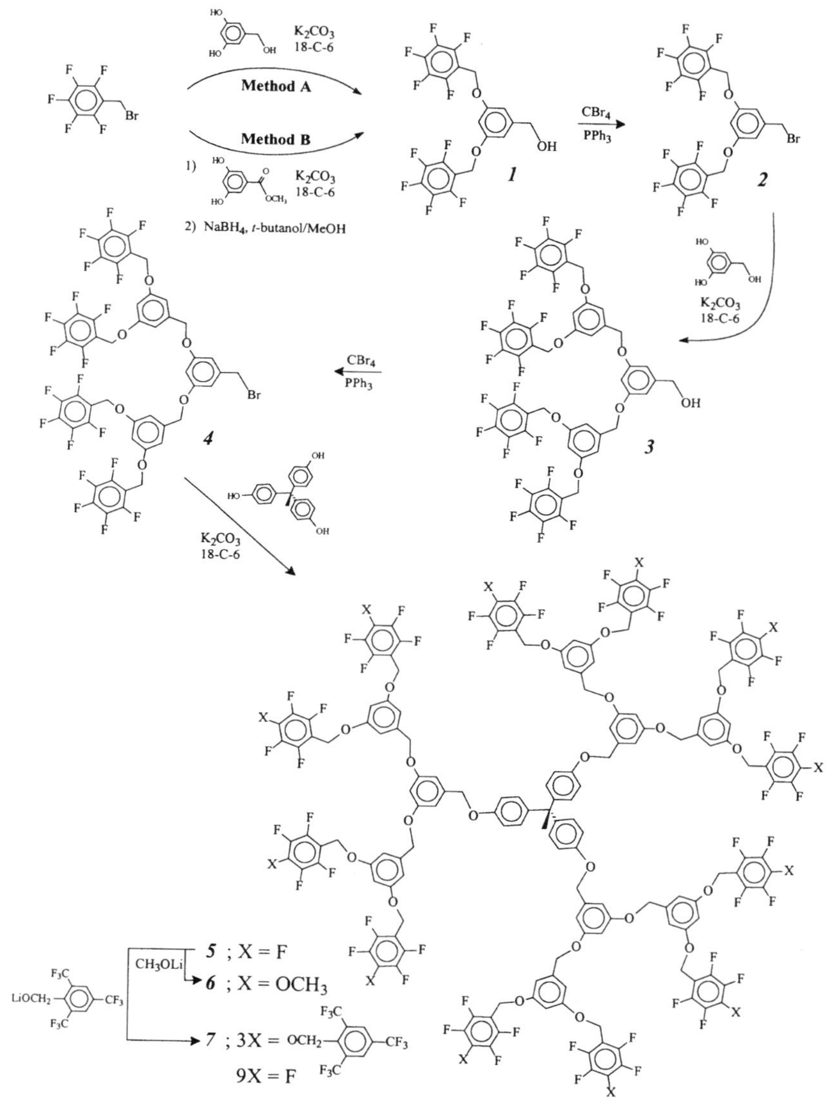

Scheme 1

the two phenolic sites to yield the first generation product **1** in 71% yield. Activation of the benzylic alcohol of **1** to an electrophilic benzylic bromide, **2**, was accomplished in 85% yield by reaction with carbon tetrabromide and triphenylphosphine. This two-step sequence of growth (alkylation) followed by activation (bromination) was then repeated to produce the second (**3,4**) and higher generation dendrimers. Alternatively, the dendritic fragments could be prepared in a three-step repetitive sequence (method B). In this case, growth was by alkylation of methyl 3,5-dihydroxybenzoate, and activation involved reduction of the methyl ester, followed by bromination. The reduction of the ester group must be done with sodium borohydride in a mixture of *tert*-butanol and methanol, to prevent hydride displacement of the *para*-fluorines. Additionally, the temperature during the alkylation reactions was kept below 55 °C and the brominations were performed at 0 °C to prevent side reactions involving the pentafluorophenyl moieties. The number of pentafluorophenyl units doubled with each generation growth, from two to four to eight to sixteen on going from generations one to two to three to four (generations three and four not shown in Scheme 1). As the generation numbers increased, the yields for the alkylation products dropped dramatically, due to competing displacement of the *para*-fluorine atoms of the growing numbers of pentafluorophenyl groups, as well as the typical steric effects experienced with the convergent growth approach. Although this may appear to have been a limitation in the synthesis, the reactivity of the pentafluorophenyl group served as a convenient method for modification of the chain end functionalities of the dendrimers to alter the properties. Additionally, as described in sections B and C, substitution of the *para*-fluorine atoms of the pentafluorophenyl groups provides a route for the preparation of hyperbranched polymers of high molecular weight in a single polymerization step.

Reactivity

In order to perform nucleophilic displacement reactions at the chain ends of the pentafluorophenyl-terminated dendrimers, the reactivity of the focal point was eliminated through coupling of the second generation dendritic bromide to the trifunctional core, 1,1,1-*tris*(4'-hydroxyphenyl)ethane, to yield **5** in 77% yield. Initial investigation of the reactivity of the dendrimer chain ends with nucleophiles involved the reaction of **5** with lithium methoxide, which was found to give complete conversion to the hexamethoxy-tetrafluorophenyl-terminated dendrimer, **6**, as observed by ^{1}H and ^{19}F NMR spectra. Appearance of the resonance for the methyl protons of the methoxy groups of **6** at 4.07 ppm and a slight upfield shift in the benzylic methylene protons of the chain end groups were observed in the ^{1}H NMR, while complete disappearance of the fluorine resonance of the *para*-fluorine atom in the ^{19}F NMR spectrum occurred along with changes in the chemical shifts of the remaining *ortho*- and *meta*-fluorines, following reaction with the methoxide.

Even though the dendritic structure provides for a high concentration of chain end groups within a small volume (in comparison to a linear structure of equivalent degree of polymerization), an increase in the density of the fluorocarbons and the introduction of alkyl fluorides was necessary to increase the fluorocarbon character of the materials. Therefore, the lithium salt of 2,4,6-*tris*(trifluoromethyl)benzyloxide, prepared by reaction of 2,4,6-*tris*(trifluoromethyl)phenyllithium with paraformaldehyde, was allowed to react with **5**. Even under forcing reaction conditions, only 25% of the chain ends of **5** could be substituted, as determined by both ^{1}H and ^{19}F NMR. It is not known whether the limited functionalization is due to steric effects and/or incompatibilities arising from fluorophobic effects.

Characterization

Standard characterization techniques included infrared spectroscopy, ^{1}H, ^{13}C, and ^{19}F NMR spectroscopy, elemental analysis, differential scanning calorimetry (DSC), and gel permeation chromatography (GPC). The expected absorbance bands were observed in the IR, which was convenient, especially for detection of the focal point functionalities following transformations from benzylic alcohol to benzylic bromide groups. NMR spectra of the three nuclei (^{1}H, ^{13}C, and ^{19}F) allowed for characterization of the structures and determination of purities. For example, side reactions involving the *para*-fluorine site were easily detected by ^{19}F NMR. In addition, both ^{1}H and ^{19}F NMR were invaluable in quantifying the extent of chain end functionalization. GPC confirmed that each of the products was a single compound of narrow polydispersity. Each of the compounds was highly soluble in common organic solvents, including chloroform, dichloromethane, tetrahydrofuran, toluene and acetone. Differential scanning calorimetry was used to measure the glass transition temperatures. As expected, the T_g increased with increasing molecular weight. The T_g for the pentafluorophenyl-terminated dendrimers were nearly 20 °C higher than those reported for their hydrocarbon analogs (Wooley, 1993). For example, compounds **1**, **3**, and **5** exhibited T_g at 4, 33 and 52 °C, while their hydrocarbon analogs were reported to give T_g at –18, 12 and 36 °C, respectively. Replacement of the *para*-fluorine atoms of **5** with methoxy groups (**6**), resulted in a decrease in the T_g from 52 to 33 °C. Introduction of the 2,4,6-*tris*(trifluoromethyl)benzyloxy groups caused further free volume increases, observed as a more substantial decrease in T_g to a transition observed at 6 °C.

The effects resulting from the incorporation of fluorine into the dendritic materials were further investigated by contact angle measurements of water on the surface of thin films. The contact angles were dependent upon the nature of the single focal point group and to a greater extent upon the composition of the multiple chain ends. Conversion of the benzylic alcohol **3** to the benzylic bromide **4** gave only a slight change in contact angle ($\theta_{\text{water, 3}} = 92°$; $\theta_{\text{water, 4}} = 93°$), and only a 2° difference in contact angle occurred upon removal of the reactive focal point group through attachment of **4** to the trifunctional core ($\theta_{\text{water, 5}} = 95°$). However, a 4° decrease in the contact angle was observed upon formation of **6** ($\theta_{\text{water, 6}} = 91°$), indicating a decrease in the hydrophobicity, upon replacement of the twelve *para*-fluorine atoms with twelve methoxy groups. Conversely, an increase in the hydrophobic nature of the material was observed with the incorporation of the 2,4,6-*tris*(trifluoromethyl)benzyloxy groups, in which a 10° increase in contact angle was observed at only 25% chain end substitution ($\theta_{\text{water, 7}} = 105°$). Furthermore, the contact angle for the non-fluorinated analog to **3** is 83°, which is less than the angles measured for all of the fluorine-containing materials.

Due to their unique architecture, dendrimers are unlike linear polymers and are unable to form extensive entanglements in the condensed phase. For the same reason, dendrimer solids exhibit very low mechanical strength, which limits the applicability of traditional methods (tensile, impact, dynamic mechanical analysis) to study the mechanical properties of dendrimer solids. Therefore, atomic force microscopy (AFM) was employed to explore the mechanical properties of dendrimers.

Atomic force microscopy (AFM) images the surface of materials by monitoring the vertical motion of an ultra-sharp (10–100 nm) tip mounted on a microcantilever while scanning laterally above the surface in an XY raster mode (Binnig, 1986). Numerous reports describe the use of scanning force microscopy to study the elastic, viscoelastic and frictional properties of materials (Mate, 1987; Burnham, 1992; Weisenhorn, 1992; Kajiyama, 1997; Friedenberg, 1996). The informative character of experiments involving tip-induced irreversible deformation of ultra-thin films has been also demonstrated (Meyers, 1992). For the purpose of the current study, the latter approach was applied with the expectation that it would best reflect the consequence of the absence of entanglements in dendrimer solids.

The method employed is a recently-developed two-step procedure, which allows for efficient testing of the behavior of a material over a range of applied forces. In the first step ("machining scan"), the AFM scan is performed over a square area (5 μm × 5 μm area, at a scanning frequency of 4.01 Hz) in the constant force mode, but with the contact force increasing in a step-wise manner at equal time intervals (20 or 25 nN steps at eight intervals). In the second step ("evaluation scan"), the contact force is minimized and a larger area scan is performed to visualize the extent of irreversible deformation produced by the probe tip (10 μm × 10 μm area at an angle of 90 degrees).

(a)

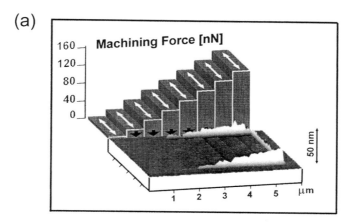

(b)

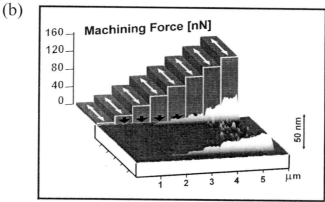

Figure 1 Atomic force microscopy micromachining experiments and resulting images for fourth generation benzyl ether dendrimers with (a) phenyl and (b) penta-fluorophenyl chain ends. The machining experiments were done by first machining with force applied by the AFM probe tip and scanning in the direction of the arrows, with increasing force from left to right, and then imaging the surface topography generated by plastic flow deformation.

The AFM images of the fourth generation dendrimers (Figure 1) demonstrate the typical micromachining behavior of dendritic materials. Micromachining of dendrimer films proceeded by displacement of material to both sides of machining-scan area and formation of rectangular depressed areas (troughs). The depths of the troughs increased with increases in the applied force. The pentafluorophenyl-substituted fourth generation benzyl ether dendrimer (with sixteen pentafluorophenyl chain ends and molecular weight of 4731.1 Da) machined more extensively at all forces ranging from

20 to 175 nN than did the non-fluorinated phenyl-terminated analog (with sixteen phenyl chain ends and a molecular weight of 3291.7 Da), for thick films of the materials (> 20 nm thickness). This behavior was somewhat surprising, given the higher T_g for the fluorinated material, and supports the expectation that the pentafluorophenyl-substituted dendrimers will perform as superior lubricating materials. Because the fluorine-containing dendrimers exhibited interesting behavior, but proved to be difficult to synthesize, accelerated routes for the preparation of similar materials were investigated.

A₂B Hyperbranched Polymers

Synthesis

Examination of the first generation dendrimer, **1**, reveals that there are two types of compatible reactive groups present in the structure. As was shown in the dendrimer functionalization reactions, the pentafluorophenyl groups are readily substituted by nucleophilic agents, in particular alkoxides. Therefore, the two pentafluorophenyl groups and the single benzyl alcohol compose the A and B functionalities required for an A₂B monomer. Formation of the reactive benzyl oxide can be done using a variety of bases, to effect the polymerization and yield the hyperbranched polymer **8**. The optimized reaction conditions for the polymerization, while avoiding side reactions, involved the addition of sodium dispersion (< 0.1 mm, 30 wt% in toluene) to **1** at 0.4 M concentration in tetrahydrofuran and heating at reflux for several days. Typical yields were *ca.* 80% following precipitation into water and purification by flash chromatography (30% hexanes/ methylene chloride as eluent), and the molecular weights ranged from *ca.* 1000 to 1,000,000 Da, depending upon the reaction conditions (Mueller, 1998).

Reactivity

A large number of pentafluorophenyl chain end groups remain throughout the macromolecule, which are available for reaction after the polymerization is complete. While the overall structure is highly branched and contains dendritic branched, dendritic end, and linear units (Scheme 2), the average repeat unit contains one tetrafluorophenyl unit, resulting from consumption of a pentafluorophenyl group to allow for polymerization, and one additional pentafluorophenyl unit. To increase the fluorocarbon character and decrease the reactivity of **8**, perfluoroalkyl chains were substituted upon the remaining pentafluorophenyl rings within the structure.

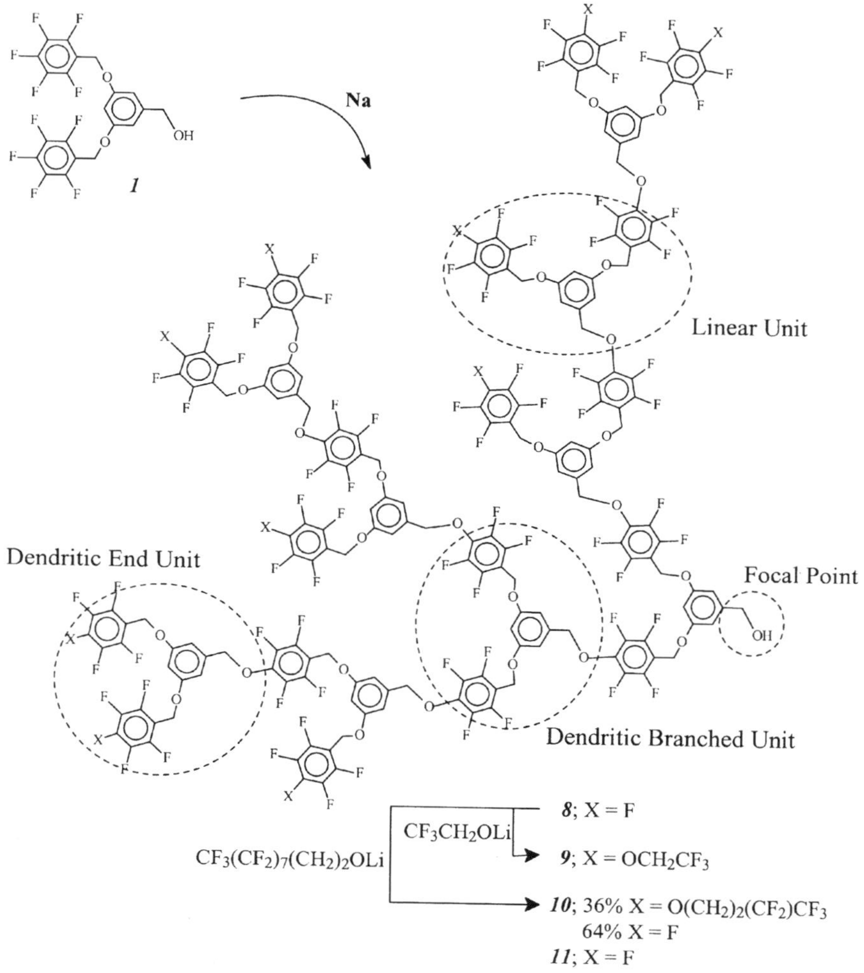

Scheme 2

The 2,2,2-trifluoroethanoxy-substituted hyperbranched polyfluorinated benzyl ether polymer **9** was synthesized by deprotonation of tri-fluoroethanol (1.5 equivalents) using *sec*-butyllithium (1.2 equivalents), followed by addition of the lithium trifluoroethoxide to **8** in THF and heating at 50 °C under argon for three days. Purification by precipitation into water to remove excess 2,2,2-trifluoroethanol, followed by silica flash chromatography twice gave **9** in 38% yield.

1H,1H,2H,2H-perfluorodecanoxy-substituted hyperbranched polyfluorinated benzyl ether polymers **10** and **11** were synthesized in the same manner
as for **9**, with the exception that the alcohol used was 1H,1H,2H,2H-
perfluorodecanol. A large excess of alcohol was needed to achieve high
substitution: 1.2 equiv. of lithium 1H,1H,2H,2H-perfluorodecanoxide
yielded 36% substitution, (**10**), while 4.9 equiv. yielded 95–100% substitution (**11**). Purification of **10** and **11** was by repeated precipitation into
hexane to remove excess alcohol and then by silica flash chromatography
twice to yield 34% and 43%, respectively. The low yields obtained were due
to isolation and purification difficulties.

Characterization

As with the dendrimer samples, all of the compounds were characterized by
standard techniques including infrared spectroscopy, ^{1}H,^{13}C, and ^{19}F NMR
spectroscopy, elemental analysis, gel permeation chromatography (GPC),
differential scanning calorimetry (DSC) and thermogravimetric analysis
(TGA). Molecular weights were measured by GPC in comparison to polystyrene standards. The number average molecular weights (M_n) of samples
having lower degrees of polymerization were also calculated by ^{1}H and ^{19}F
NMR analysis. The hyperbranched polymers were thermally stable up to
305°C in air, as determined by TGA.

Contact angles of water (θ_{water}) and hexadecane ($\theta_{hexadecane}$) were measured by the sessile drop method (Neumann, 1979) on the surfaces of films
of **8** and its derivatives, **9**, **10**, and **11**. The hydrophobicity of **1** was not high,
$\theta_{water, 1} = 69°$, most likely due to the OH functionality. After polymerization
of **1**, the concentration of OH groups decreased and the contact angle of
water upon **8** was 96°. The addition of trifluoroethoxy groups (**9**) did not
substantially change the hydrophobicity of the material ($\theta_{water, 9} = 99°$).
However, substitution of only 36% of the chain ends with 1H,1H,2H,2H-
perfluorodecanoxy groups (**10**) increased the contact angle to
$\theta_{water, 10} = 120°$. The contact angle of water on poly(tetrafluoroethylene)
(PTFE) was 115°, under our conditions. The contact angle did not change
further when 100% of the *para*-fluorines were substituted with
1H,1H,2H,2H-perfluorodecanol ($\theta_{water, 11} = 120°$).

Each of the materials, **1**, **8** and **9**, was not very lipophobic, with
$\theta_{hexadecane} = 18°$, 21° and 14°, respectively. The highest lipophobicity was seen
for **10** and **11**, and these contact angles ($\theta_{hexadecane, 10 \text{ and } 11} = 62°$ and 67°,
respectively) were considerably higher than the contact angle of PTFE
($\theta_{hexadecane} = 42°$) under these conditions.

Atomic force microscopy (AFM) was used to study the morphology and
surface properties of the materials. Surfaces of spin-coated films of **8** and **11**

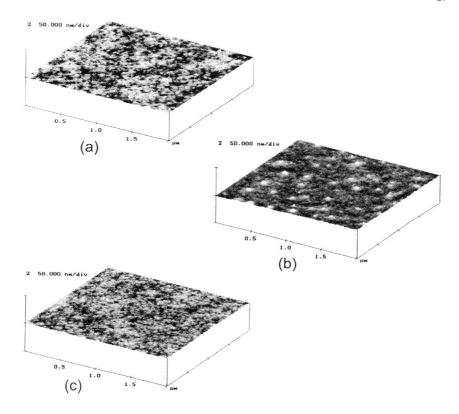

Figure 2   AFM images illustrating the surface topographies of (a) **8**, (b) **10**, and (c) **11**. Both of the samples with homogeneous substitution of the chain ends (**8** and **11**) appear to have microscopically rough surface features, whereas the partially perfluoroalkyl-substituted material (**10**) displays phase separated domains.

(Figure 2(a) and 2(c)) exhibited uniform fractal morphology, characteristic for glassy polymers with root-mean-square surface roughness in the range 0.17–0.26 nm (Kowalewski, 1997). However, films of **10**, in which only 36% of available sites were substituted with fluoroalkyl chains, exhibited markedly different morphology (Figure 2(b)) with distinct 4 ± 1 nm protrusions (average height based on bearing analysis). The lateral size of the protrusions did not exceed 100 nm. Such morphology suggests phase separation between the fractions with different extents of substitution. Due to similar T_g for the substituted and unsubstituted polymers, this phase separation was not detected by DSC.

The surface properties of **8**, **10**, and **11** were studied with lateral force experiments using contact mode AFM (Binnig, 1986; Meyer, 1988, 1990).

The coefficients of friction were measured as 0.065 and 0.028 and the adhesive forces were 93 and 38 nN for **8** and **11**, respectively. These data demonstrate more than a two-fold decrease of the coefficient of friction, μ, and of the adhesive force, F_a, upon substitution with perfluoroalkyl chains onto the initial hyperbranched polymer (**8** *vs.* **11**). Films of **10** (partially substituted) exhibited mixed characteristics depending on the region of the film; the film surface gave $\mu = 0.035$ and the elevated domains gave $\mu = 0.068$. Comparison of the coefficients of friction for **8**, **10**, and **11** suggests that the phase-separated domains in **10** are composed of mainly the parent pentafluorophenyl-terminated polymer, while the areas between the phase-separated domains (the film surface) contain the fluoroalkyl-chain substituted polymer. It is worth noting that the coefficient of friction, μ, and the adhesive force, F_a, observed for **11** were significantly lower in comparison with PTFE, despite substantially higher fluorine content for PTFE. We propose that this points to a very effective modification of surface properties when substitution is performed with relatively short perfluoroalkyl chains. Such effectiveness may stem from a large fraction of chain-end CF_3 groups introduced into the material and from high molecular mobility of the fluoroalkyl chains when attached to the chain ends of hyperbranched polymers.

A_4B Hyperbranched Polymers

Synthesis

To conveniently prepare a high molecular weight polymer with an increase in the degree of branching, hyperbranched polymers were then prepared by polymerization of the second generation pentafluorophenyl-terminated dendrimer, **3** (Stefanescu, 1997). The second generation structure is an A_4B monomer, as it contains four pentafluorophenyl groups and a single benzyl alcohol. Polymerization proceeds by deprotonation of the benzyl alcohol, followed by nucleophilic displacement of any of the reactive *para*-fluorine atoms found in **3** by the benzyloxy anion of another monomer or polymer molecule. As macromolecular growth proceeds, the number of pentafluorophenyl reactive sites increases ($\#A = 3n + 1$, where $n =$ degree of polymerization), while the single benzyloxy group is maintained. However, under certain reaction conditions, intramolecular reaction was observed to lead to cyclic products (Gooden, 1998).

The polymerization conditions were similar to, but more mild than, those used in the A_2B polymerization. Solutions of 0.1 to 0.3 M of **3** in tetrahydrofuran were treated with 1.5 to 2 equiv. of Na suspension in toluene (30 wt%, < 0.1 mm particle size), and allowed to react under N_2 at room

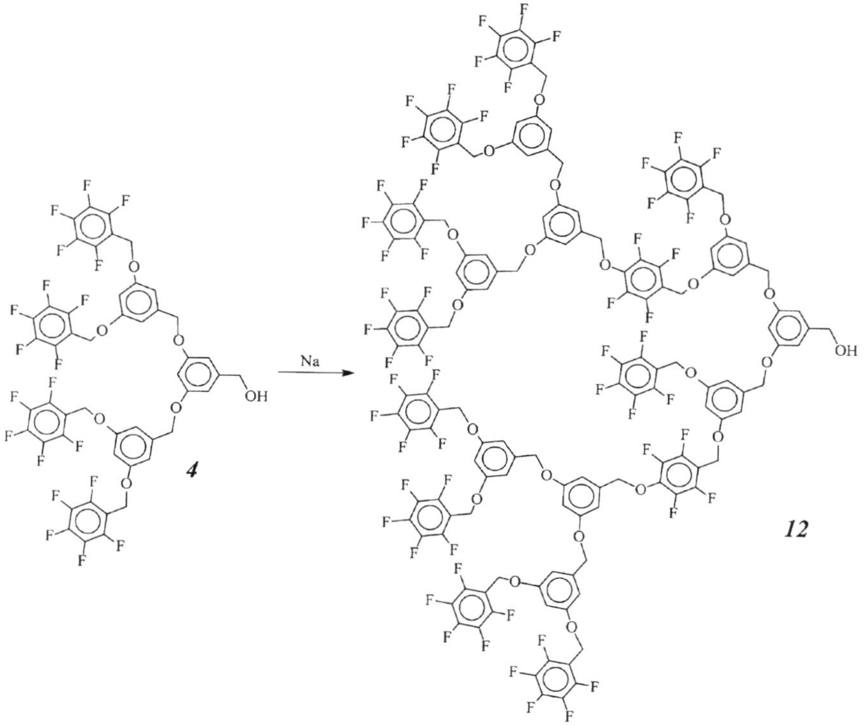

Scheme 3

temperature. The polymerization was allowed to proceed for 2 to 30 hours, while being monitored by gel permeation chromatography. The polymerization was then quenched in water, and the precipitated crude polymer was isolated by centrifugation. The crude polymer was then solubilized in a minimum amount of tetrahydrofuran and filtered through silica gel to remove any polar contaminants. The A_4B hyperbranched polymer, **12**, was obtained in typically 65% yield as a white solid. The structure illustrated in Scheme 3 is a trimer, which is meant only to represent the general polymer composition; molecular weights ranged typically from 2000 to 20000 Da.

Characterization

The A_4B hyperbranched polymer gave spectroscopic, thermal, and surface characterization that was similar to the A_2B hyperbranched polymer, due to the similarities in functional and elemental constitution. However, there are differences in the structural repeat unit, degree of branching, fluorine content and fluorine distribution between **12** and **8**. For example, the %F (a/a) is 22% and 20% for hexamers of **8** and **12**, respectively, and **12** contains a higher proportion of pentafluorophenyl : tetrafluorophenyl groups; these differences are observed as differences in relative intensities in the ^{19}F NMR spectra. In addition, a calculated degree of branching (from growth) of 44% for **12** gives an effective branching degree of 83%, due to the built-in branching units (from the second generation monomer repeat structure). More elaborate AFM experiments to probe differences in micromachining behavior are in progress.

CONCLUSIONS

Highly-branched polyfluorinated polymers of varying degrees of branching and bearing different chain end functionalities were prepared to investigate the effects of the combined branching and fluorocarbon content upon reduction of the intermolecular interactions and the resulting properties of the materials. The pentafluorophenyl group proved to be a useful functionality that was stable during construction of the materials, but then offered a reactive site for post-polymerization modification by nucleophilic aromatic substitution selectively upon the *para*-fluorine atoms. The dendrimers are well-defined structures with a branching site at each repeat unit and all chain ends structurally emanating from a central core, but are not trivial to prepare in high yields and of high molecular weight due to the stepwise synthesis. Polymerization of the low generation dendrimers leads to high molecular weight hyperbranched polymers in convenient one-step processes. The hyperbranched polymer prepared from the first generation dendrimer (A_2B monomer) is expected to contain the statistical 50% branching and the development of a degradative scheme to determine this is in progress. Polymerization of the second generation dendrimer (A_4B monomer) builds branching into each repeat unit so that the end polymer should possess effective branching degrees between 50 and 100% and the materials should exhibit intermediate properties between dendrimers and A_2B hyperbranched polymers. Each of the materials forms a brittle film that exhibits hydrophobicity and lipophobicity, these properties are enhanced by the incorporation of perfluoroalkyl chain end groups. The materials have low coefficients of friction, low adhesive forces and are micromachined at nN forces by an atomic force microscope probe tip. At present, the nature of the fluorocarbon functionality appears to dominate the properties of the materials, although further elucidation of the differences resulting from the differences in branching degrees is in progress.

ACKNOWLEDGMENTS

Financial support for this work by a U.S. Army Research Office Young Investigator Award (DAAH04-96-0158, K.L.W.), a U.S. National Science Foundation National Young Investigator Award (DMR-9458025, K.L.W.) and the U.S. Office of Naval Research (T.K.) is gratefully acknowledged.

REFERENCES

Binnig, G., Quate, C.F. and Gerber, C. (1986). "Atomic Force Microscope", *Phys. Rev. Lett.*, **56**, 930.

Brady, R.F. (1990). "Fluoropolymers", *Chem. in Britain*, 427.

Burnham, N.A. and Colton, R.J. (1989). *J. Vac. Sci. Technol.*, **A7**, 2906.

Chambers, R.D. (1973). *Fluorine in Organic Chemistry*; Interscience Monographs on Organic Chemistry, Olah, G.A. (Ed.), John Wiley & Sons: New York.

Fréchet, J.M.J. (1994). "Functional Polymers and Dendrimers: Reactivity, Molecular Architecture, and Interfacial Energy", *Science*, **263**, 1710.

Friedenberg, M.C. and Mate, C.M. (1996). *Langmuir*, **12**, 6138.

Gangal, S.V. (1989). *Encyclopedia of Polymer Science and Engineering*, Mark, H.F., Bikales, N.M., Overberger, C.G. and Menges, G. (Eds), Wiley Interscience: New York. Vol. 16, pp. 677–9.

Gooden, J.K., Gross, M.L., Mueller, A., Stefanescu, A.D. and Wooley, K.L. (1998). "Cyclization in Hyperbranched Polymer Syntheses: Characterization by MALDI-TOF Mass Spectrometry", *J. Am. Chem. Soc.*, **120**, 10180.

Hawker, C.J., Malmström, E.E., Frank, C.W. and Kampf, J.P. (1997). "Exact Linear Analogs of Dendritic Polyether Macromolecules: Design, Synthesis, and Unique Properties", *J. Am. Chem. Soc.*, **119**, 9903.

Kajiyama, T., Tanaka, K. and Takahara, A. (1997). *Macromolecules*, **30**, 280.

Kowalewski, T. and Schaefer, J. (1997). *ACS Polym. Matl. Sci. Eng. Prepr.*, **76**, 215.

Mate, C.M., McClelland, G.M., Erlandsson, R. and Chiang, S. (1987). *Phys. Rev. Lett.*, **59**, 1942.

Meyer, G. and Amer, N.M. (1988). "Novel Optical Approach to Atomic Force Microscopy", *Appl. Phys. Lett.*, **53**, 1045.

Meyer, G. and Amer, N.M. (1990). "Simultaneous Measurement of Lateral and Normal Forces with an Optical-Beam-Deflection Atomic Force Microscope", *Appl. Phys. Lett.*, **57**, 2089.

Meyers, G.F., DeKoven, B.M. and Seitz, J.T. (1992). "Is the Molecular Surface of Polystyrene Really Glassy?", *Langmuir*, **8**, 2330.

Mueller, A., Kowalewski, T. and Wooley, K.L. (1998). "Synthesis, Characterization and Derivatization of Hyperbranched Polyfluorinated Polymers", *Macromolecules*, **31**, 776.

Neumann, A.W. and Good, R.J. (1979). *Surface and Colloid Science*, **11**, 31.

Advances in Dendritic Macromolecules, Vol. 1 & 2; Newkome, G.R., Ed.; JAI Press: Greenwich, CT, **1994**, **1995**.

Smart, B.E. (1994). *Organofluorine Chemistry, Principles and Commercial Applications*, Banks, B.E., Smart, B.E. and Tatlow, J.C. (Eds), Plenum Press: New York and London. pp. 57–88.

Stefanescu, A.D. and Wooley, K.L. (1997). "Hyperbranched Polyfluorinated Polymers from AB4 Monomers", *ACS Polym. Matl. Sci. Eng. Prepr.*, **77**, 218.

Tomalia, D.A., Naylor, A.M. and Goddard, W.A., III (1990). "Starburst Dendrimers: Control of Size, Shape, Surface chemistry, Topology and Flexibility in the Conversion of Atoms to Macroscopic Materials", *Angew. Chem.*, **102**, 119.

Weisenhorn, A.L., Maivald, P., Butt, H. -J. and Hansma, P.K. (1992). *Phys. Rev. B.*, **45**, 11226.

Wooley, K.L., Klug, C.A., Tasaki, K. and Schaefer, J. (1997). "Shapes of Dendrimers from Rotational-Echo Double-Resonance NMR", *J. Am. Chem. Soc.*, **119**, 53.

Wooley, K.L., Hawker, C.J., Pochan, J.M. and Fréchet, J.M.J. (1993). "Physical Properties of Dendritic Macromolecules: A Study of Glass Transition Temperature", *Macromolecules*, **26**, 1514.

3. ELECTRONICALLY CONTROLLABLE HIGH SPIN SYSTEMS REALIZED BY SPIN-POLARIZED DONORS

TADASHI SUGAWARA, HIROMI SAKURAI and AKIRA IZUOKA

Department of Basic Sciences, The University of Tokyo, 3-8-1 Komaba, Meguro-ku, Tokyo 153-8902, Japan

INTRODUCTION

Molecular magnetism is a young research area which has grown rapidly in the last two decades.[1] An advantage of molecular magnetism is that one can design and prepare spin sources at the molecular level.[2] Since many unconventional spin systems are now available by organizing organic radicals, the next stage of advancement in molecular magnetism is to construct a dynamic spin system, the spin states of which can be controlled by means of photoirradiation (Figure 1(a)), electron attachment or detachment (Figure 1(b)), transportation of electrons, or modulation of crystal lattices (Figure 1(c)). Development of such dynamic spin systems can be achieved by taking advantage of flexible organic molecules carrying electron spins.

In this chapter, the authors describe design and preparation of radical donors which afford a ground state triplet cation diradical upon one-electron oxidation, and discuss the characteristic electronic structure of such radical donors based on the perturbational molecular orbital (*PMO*) method.

(a)

Photochemical control

(b)

Electronic control

(c)

Lattice control

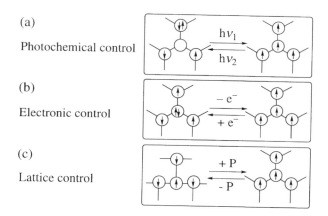

Figure 1   Dynamic spin system controllable by external stimuli.

The spin distribution of these radical donors turns out to be polarized significantly and they can be called *spin-polarized* donors. Finally, the possibilities of constructing dynamic spin systems using such *spin-polarized* donors are documented.

1. SELF-ASSEMBLING ORGANIC SPIN SYSTEM

1.1. Magnetic Interactions between Organic Radicals

Since magnetism is a bulk property, one has to take into account intermolecular magnetic interactions among organic radicals. When organic radicals assemble, the mutual magnetic interaction is usually very weak, and they behave paramagnetically (Figure 2, left). Even when they interact,

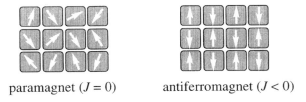

paramagnet ($J = 0$)   antiferromagnet ($J < 0$)

Figure 2 Intermolecular magnetic interactions in an assembly of organic radicals. J denotes an intermolecular exchange interaction.

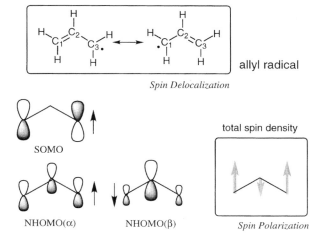

Figure 3 Spin delocalization and spin polarization in allyl radical.

(a) (b) (c)

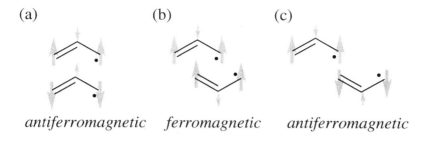

antiferromagnetic   ferromagnetic   antiferromagnetic

Figure 4 Correlation between relative orientations and intermolecular magnetic interactions of allyl radicals.

unpaired electrons tend to be paired, resulting in antiferromagnetic materials (Figure 2, right). The question is, how can we align electron spins of organic radicals in parallel?

Organic π-radical, in general, has a characteristic electronic structure. In an allyl radical, for example, an α-spin in SOMO (singly occupied molecular orbital) distributes on C1 and C3 carbons (*spin delocalization*), and the spin distribution of SOMO influences those of α- and β-spins in NHOMO (next highest occupied molecular orbital). While the distribution of the α-spin is unaffected, that of the β-spin trends to avoid residing on C1 and C3. As a result, a negative spin density is resulted on C2. This modulation in the spin distribution is called *spin polarization*.

The next stage is to consider the intermolecular magnetic interaction of allyl radicals in various orientations. When allyl radicals are aligned in the arrangement (a) or (c) in Figure 4, the overall intermolecular interaction becomes antiferromagnetic. This is because the local intermolecular interactions between spin-distributing carbon atoms are antiferromagnetic based on the kinetic exchange interaction. The arrangement (b) in Figure 4 leads to a net ferromagnetic interaction by the same token. Consequently, it may be clear that the intermolecular magnetic interaction depends heavily on the relative orientations in a molecular assembly of π-radicals.[3] Magnetic interactions observed in cyclophane dicarbenes with various substituting patterns proved the above mechanism of the spin ordering experimentally.[4]

1.2. Construction of Supramolecular Spin System

Under such circumstances, it is of great significance to regulate crystal structures of organic radicals so as to exhibit desirable magnetic properties. The method for controlling the crystal structure of organic radicals proposed here involves preparation of organic radicals equipped with an *orientation controlling site* (*OCS*).[5] An *OCS* is a functional group which

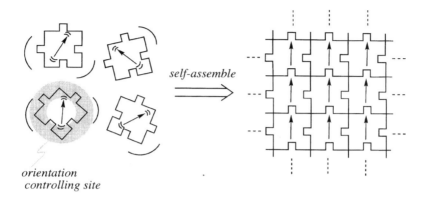

*orientation
controlling site*

Figure 5 Construction of a ferromagnetic self-assembly in the use of organic radicals equipped with an orientation controlling site.

plays a significant role of self-assembling organic radicals by means of intermolecular interactions, such as electrostatic interactions, hydrogen bonds, or charge transfer interactions, *etc.* (Figure 5). Until now, various spin systems have been constructed using *OCS* containing organic radicals.

Representative examples are shown in Figure 6, such as molecular-based ferromagnets,[6] a ferrimagnetic spin system,[7] a quasi-one-dimensional ferromagnet,[8] a weak ferromagnet,[9] and a Kagomé lattice,[10] which can not be available in inorganic materials, have been prepared. A nitronyl nitroxide group or a verdazyl group carries an unpaired electron and the other part of the molecule, such as a *p*-hydroquinonyl, *m*-dinitrophenyl, chlorophenyl, or pyridyl group, behaves as an *OCS*. One of the successful cases for arranging π-radicals to exhibit ferromagnetic interaction is a hydrogen-bonded ferromagnet,[6a,b] abbreviated as **HQNN** (a nitronyl nitroxide derivative substituted by *p*-hydroquinone), which utilizes the hydrogen-bonding ability of the hydroquinone moiety as an *OCS*. Since the arrangement of crystal **HQNN** satisfies the ferromagnetic requirement, the crystal exhibits a ferromagnetic phase transition at 0.5 K, showing a hysteresis loop in its magnetization curve.

2.   DESIGN AND PREPARATION OF "SPIN-POLARIZED" DONORS

2.1.   Spin-Polarized Donor as a Building Block of Dynamic Spin Systems

In order to construct a "dynamic spin system controllable by external stimuli", a radical donor, which affords a cation diradical of a ground state

3-D Ferromagnet

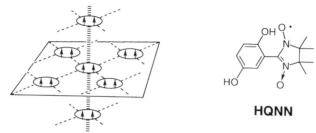

HQNN

1-D Ferrimagnetic Chain

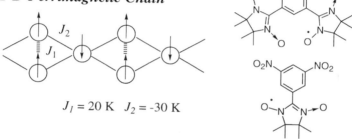

$J_1 = 20$ K $J_2 = -30$ K

Quasi-1-D Ferromagnet

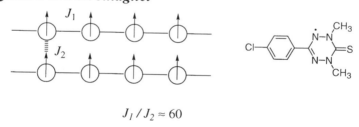

$J_1 / J_2 \approx 60$

"Kagomé" Structure

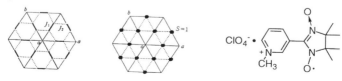

Figure 6   Various molecular spin systems.

triplet spin multiplicity upon one-electron oxidation, has been designed. This type of donor may be called a *spin-polarized* donor because of its characteristic electronic structure shown in section 2.4. The design of such a *spin-polarized* donor is described in the next section.

Based on a concept of the *spin-polarized* donor, the following novel spin systems may be constructed:

a) High spin cation polyradical, the spin multiplicity of which can be controlled by a redox process (Figure 7(a)): While the spin quantum number of polyradical donor is $S = 1/2$ in the neutral state, it is converted to $S = (n + 1)/2$ (n is a number of radical sites attached to the donor core) upon one-electron oxidation.

b) Organic ferrimagnet (Figure 7(b)): When an organic acceptor and a *spin-polarized* donor are stacked alternately, ferrimagnetic interaction can be realized.

c) Organic ferromagnetic metal (Figure 7(c)): When *spin-polarized* donors form segregated stacks with mixed valence states, conduction electrons may align local spins on the radical sites ferromagnetically.[11]

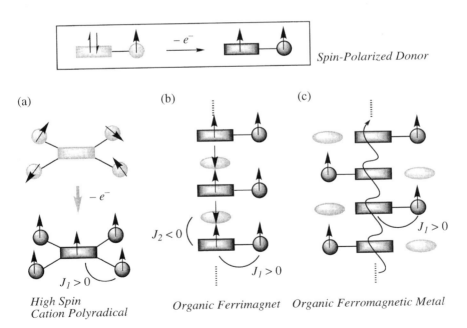

Figure 7   Schematic drawing of spin systems composed of spin-polarized donors: (a) High spin cation polyradical, (b) Organic ferrimagnet, (c) Organic ferromagnetic metal.

2.2. Molecular Design of the Spin-Polarized Donor as a Hetero-Analogue of TMM

Before describing a *spin-polarized* donor in detail, let us return to the origin of intramolecular ferromagnetic coupling in trimethylenemethane (**TMM**).[12] When a nodal carbon of an allyl radical is substituted by a methyl radical, a ground state triplet diradical, **TMM**, is obtained (Figure 8). A connection pattern in the π-conjugation of **TMM** is called cross-conjugation in a valence-bond description. The ferromagnetic coupling between two unpaired

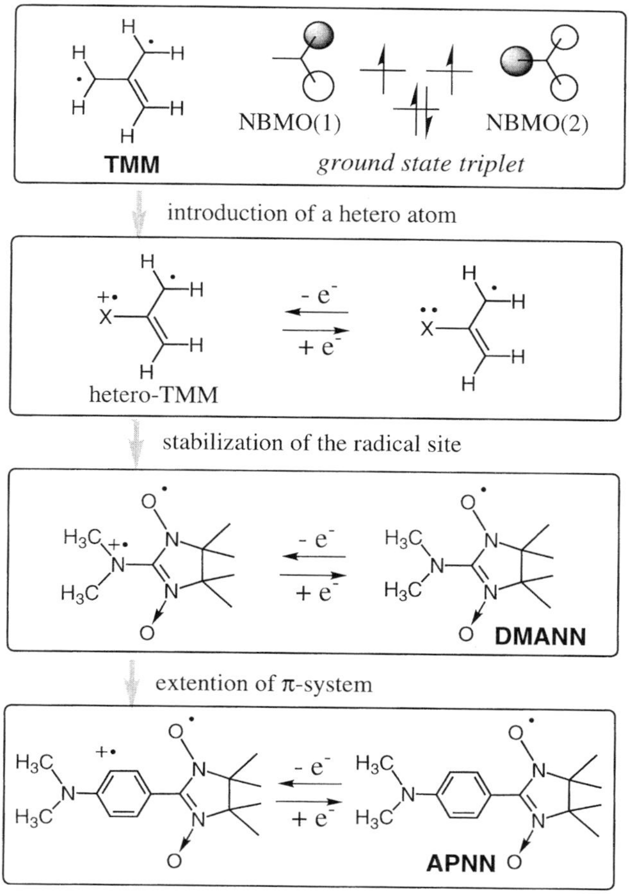

Figure 8 Design of a radical donor affording a triplet cation diradical upon one-electron oxidation.

electrons in **TMM** is significantly large ($\Delta E_{TS} = 16\,kcal/mol$ *expl.*[13]). The origin of such a large exchange interaction is derived from the nature of the degenerated non-bonding molecular orbitals (NBMOs) of **TMM**. The coefficients of NBMOs (*cf.* Figure 8) are shared by two carbon atoms of **TMM**. This means that two electrons have a chance to come closer, especially when they have opposite spins, and they suffer from an excess exchange interaction of $2K_{ij}$, where K_{ij} is an exchange integral between electrons of i and j, compared with the case where the electrons are of the triplet spin-correlation. As a result, the triplet state of **TMM** is significantly more stable than the singlet state.

If a hetero atom with a lone pair of electrons is introduced to the nodal carbon of allyl radical, the resulting radical gives rise to a hetero-analogue of **TMM**, when it is singly oxidized upon one-electron oxidation. Since the electronic structure of the resulted cation diradical resembles that of **TMM**, its ground state is considered to be triplet. In order to enhance the kinetic stability of the allyl radical part, it is replaced with nitronyl nitroxide (NN), which can be regarded as a stable hetero-analogue of allyl radical. According to the above chemical modifications, dimethylamino nitronyl nitroxide, **DMANN**, and dimethylaminophenyl nitronyl nitroxide, **APNN**, were prepared. These radical donors are expected to afford ground state triplet cation diradicals upon one-electron oxidation.

2.3. Generation of a Ground State Triplet Cation Diradical by Means of a Redox Process

Oxidation potentials of **DMANN**, **APNN** and those of reference compounds are determined by cyclic voltammetry as summarized in Table 1. The oxidation potential of **DMANN** is 0.53 V, while the oxidation potentials of the reference compounds, nitronyl nitroxide (H-NN) and dimethylamine, are 0.82 and 1.27 V, respectively. The first oxidation potential of **APNN** (0.64 V) is also lower than those of the parent compounds (*cf.* dimethylaniline 0.82 V). The lower oxidation potentials of these radical donors compared with those of the parent compounds suggest a significant electronic interaction between the donor site and the radical site.

Table 1   Oxidation potentials of radical donors and reference compounds. (vs. Ag/AgCl, in 0.1M n-Bu$_4$N.ClO$_4$/CH$_3$CN)

Compounds	$E_{1/2}$/ V	Compounds	$E_{1/2}$/ V	
		H-NN	0.82	
Me$_2$N-H	1.27 (irr.)	Me$_2$N-NN (**DMANN**)	0.53	
Me$_2$N-Ph	0.82 (irr.)	Me$_2$N-Ph-NN (**APNN**)	0.64	1.31 (irr.)

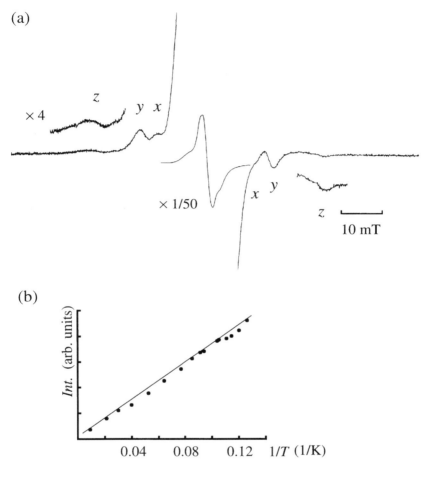

Figure 9 (a) ESR spectrum of **DMANN**⁺˙ in THF glass at 8 K, (b) Temperature dependence of the triplet signal intensity of **DMANN**⁺˙.

Oxidized species of the radical donors were generated by the addition of excess iodine to a THF solution at room temperature. Figure 9(a) shows a characteristic triplet ESR spectrum of the oxidized **DMANN** in a frozen matrix.[14] The triplet ESR spectrum can be interpreted by the following Hamiltonian.

$$H = g\beta \mathbf{H} \cdot \mathbf{S} + \mathbf{S} \cdot \mathbf{D} \cdot \mathbf{S}$$
$$= g\beta \mathbf{H} \cdot \mathbf{S} + D[S_z^2 - S(S+1)/3] + E(S_x^2 - S_y^2)$$

The zero-field splitting parameters calculated from the resonance fields were $|D| = 0.0276\,\text{cm}^{-1}$ and $|E| = 0.0016\,\text{cm}^{-1}$. Since the signal intensity of the ESR spectrum of the triplet species obey the Curie law in the temperature range of 8–100 K (Figure 9(b)), the triplet is considered to be the ground state of **DMANN$^{+\bullet}$**. The similar tendency was obtained for **APNN**. The zero-field splitting parameter for **APNN$^{+\bullet}$** are $|D| = 0.0272\,\text{cm}^{-1}$, and $|E| = 0.0018\,\text{cm}^{-1}$.

The $|D|$ parameter of **DMANN$^{+\bullet}$** is smaller than the estimated value based on a point-dipole approximation, and it is similar to that of **TMM** ($|D| = 0.024\,\text{cm}^{-1}, |E| < 0.001\,\text{cm}^{-1}$).[15] The tendency may be rationalized by the presence of a negative spin density induced at the α-carbon of nitronyl nitroxide as pointed out in the case of **TMM**.[16] The similarity of $|D|$ values recognized between **DMANN$^{+\bullet}$** and **TMM** is supporting evidence that the electronic structure of **DMANN$^{+\bullet}$** is close to that of **TMM**.

2.4. Characteristics of the Electronic Structure of Triplet Cation Diradical Derived from Spin-Polarized Donors

The electronic features of a radical donor can be interpreted by a perturbational molecular orbital method,[14] provided that the radical donor is composed of a donor site and a radical site, and that they interact mutually. The electronic structure of such an radical donor may be represented by either (a) or (b) as shown in Figure 10.

In case (a), one-electron oxidation will remove the electron from a singly occupied molecular orbital (SOMO) of the radical donor to afford a closed-shell cationic species. This electronic structure is, of course, unsuitable for generating a triplet cation diradical. In case (b), the highest occupied molecular orbital (*homo*) of the donor site interacts with the next highest molecular orbital (*nhomo*) of the radical site, not interacting with the singly occupied molecular orbital (*somo*) of the radical site, because of the symmetry mismatch of the relevant partial molecular orbitals. This electronic interaction raises the energy level of HOMO of the radical donor. As a result, HOMO is placed above SOMO. Such an exotic electronic structure can be maintained if the on-site Coulombic repulsion of SOMO is larger than the orbital energy difference (ΔE) between HOMO and SOMO.

These requirements are usually satisfied by nitronyl nitroxide (NN) derivatives substituted with a reasonably good donor group, because *somo* of NN is localized on the nitronyl nitroxide group consisting of electronegative atoms, and the symmetry of *somo* is a_2, which usually does not match with the symmetry of the π-orbital of the donor unit. The lower oxidation potential of **DMANN** and **APNN** compared with those of the parent compounds, supports the presence of the electronic interaction

(a)

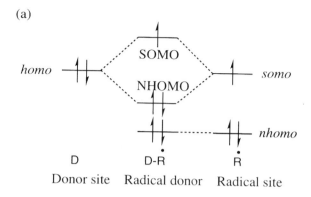

(b)

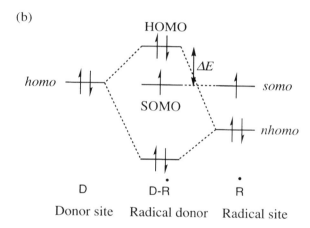

Figure 10 Schematic drawing for the electronic configuration of a radical donor: (a) *homo* of the donor site interacting with *somo* of the radical site; (b) *homo* of the donor site interacting with *nhomo* of the radical site.

between the donor site and the radical site as discussed above. If SOMO and SOMO′, the latter of which is derived from HOMO upon one-electron oxidation, and SOMO are of space-sharing types each other, two unpaired electrons residing in SOMO and SOMO′ should interact ferromagnetically to afford a triplet cation diradical.

According to the semi-empirical molecular orbital calculation (PM3/ UHF) of **DMANN**, an α-spin, residing in SOMO, causes a large spin-polarization to π-spins in HOMO (Figure 11). As a result, the energy level of the β-spin of HOMO becomes higher than that of the α-spin due to the

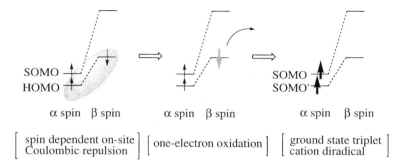

Figure 11 Electronic structure of a *spin-polarized* donor, **DMANN**, and of a triplet cation diradical generated upon one-electron oxidation in a UHF description.

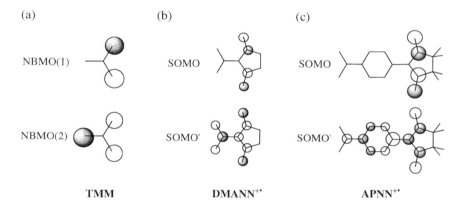

Figure 12 Coefficients of singly occupied molecular orbitals for (a) **TMM**, (b) **DMANN⁺·** and (c) **APNN⁺·** calculated by PM3/RHF method.

spin-dependent on-site Coulombic repulsion. Thus, it is not strange that the one-electron oxidation removes the β-spin of HOMO to afford a ground state triplet species. Such a radical donor can be called a *spin-polarized* donor.

In order to rationalize the triplet ground state of the cation diradical, a PM3/RHF calculation was also performed on the cation diradical of **DMANN**. The two singly occupied molecular orbitals of **DMANN⁺·** are regarded as NBMOs of alternate hydrocarbons, although the two energy levels are not degenerate. Coefficients of the singly occupied molecular orbitals of **DMANN⁺·** are displayed in Figure 12(b). As seen in Figure 12(b), the coefficients of SOMO is localized on nitronyl nitroxide, while those of SOMO′, which is derived from the one-electron oxidation of

HOMO, spread over the entire molecule, sharing the coefficients of atomic orbitals of the nitronyl nitroxide group with SOMO. As far as the distribution of the coefficients of SOMO and SOMO$'$ is concerned, it resembles closely that of the two degenerate NBMOs of **TMM** (Figure 12(a)). The cation diradical, **DMANN$^{+\bullet}$**, therefore, should exist as a ground state triplet species. In fact, a PM3/UHF calculation reveals that the triplet state of **DMANN$^{+\bullet}$** is more stable than the singlet.[14] The relevant MOs of **APNN$^{+\bullet}$** are also shown in Figure 12(c).

3. LOW SPIN-HIGH SPIN CONVERSION OF "SPIN-POLARIZED" POLYRADICAL DONORS

If the concept of a *spin-polarized* donor is applied to an extended π-system carrying plural radical sites, novel spin systems are expected to be constructed.[17] Since the intramolecular magnetic interaction between these radical sites attached to the π-donor core is considered to be negligibly small ($J_1 \approx 0$) in the neutral state, a spin quantum number of the *spin-polarized* polyradical donor is assumed to be $S = 1/2$ (Figure 13, left). When the donor core is singly oxidized, the resulted unpaired electron delocalizes over the entire molecule and align the fluctuating local spins ferromagnetically. Consequently a ground state high spin cation polyradical ($S = (n + 1)/2$) should be generated (Figure 13, right). A ferromagnetic coupling in these polyradical donors is derived from the exchange interaction between space-sharing SOMO$'$ and SOMOs; the former is generated from HOMO through one-electron oxidation, and the latters are localized on the individual radical sites. Therefore the intramolecular ferromagnetic coupling (J_1) should be reasonably large.

Along this guideline, a tetraphenylphenylenediamine or a triphenylamine skeleton was chosen as a donor core, and the core was substituted by two or three nitronyl nitroxide groups (**PDNN$_2$,TPANN$_3$**), respectively.

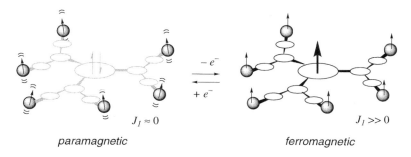

$$J_1 \approx 0 \qquad\qquad J_1 \gg 0$$

paramagnetic   *ferromagnetic*

Figure 13 Low spin-high spin conversion of polyradical donor.

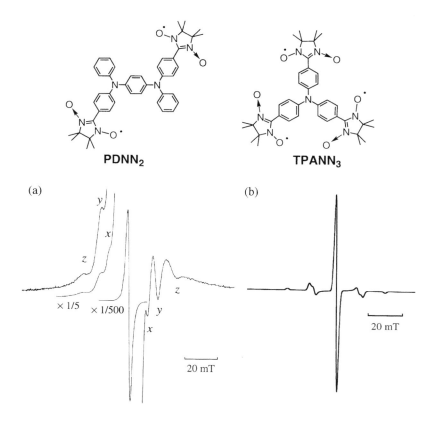

(a)   (b)

Figure 14 (a) ESR spectrum of oxidized **PDNN₂** in THF at 8 K and (b) simulated spectrum for the $S = 3/2$ species.

In order to examine the intramolecular magnetic interaction of a polyradical donor in the neutral state, temperature dependence of the magnetic susceptibility of **PDNN₂** was measured in the temperature range of 2–250 K, and the doublet ground state for **PDNN₂** was confirmed. The similar result was obtained for **TPANN₃**.

Oxidation of **PDNN₂** was carried out in THF at room temperature by the addition of excess iodine. Figure 14(a) shows the ESR spectrum of the oxidized species of **PDNN₂** measured at 8 K in a frozen matrix. The signal of the high spin species is best assigned to a quartet species based on the spectral simulation (Figure 14(b)). The $|D|$ and $|E|$ values are calculated to be 0.0134 and 0.0005 cm^{-1}, respectively. The temperature dependence of the signal intensity suggests that the quartet is the ground state of the oxidized species (Figure 15(a)) or that the quartet state lies above the lower spin states by less than *ca.* 10 cal/mol.

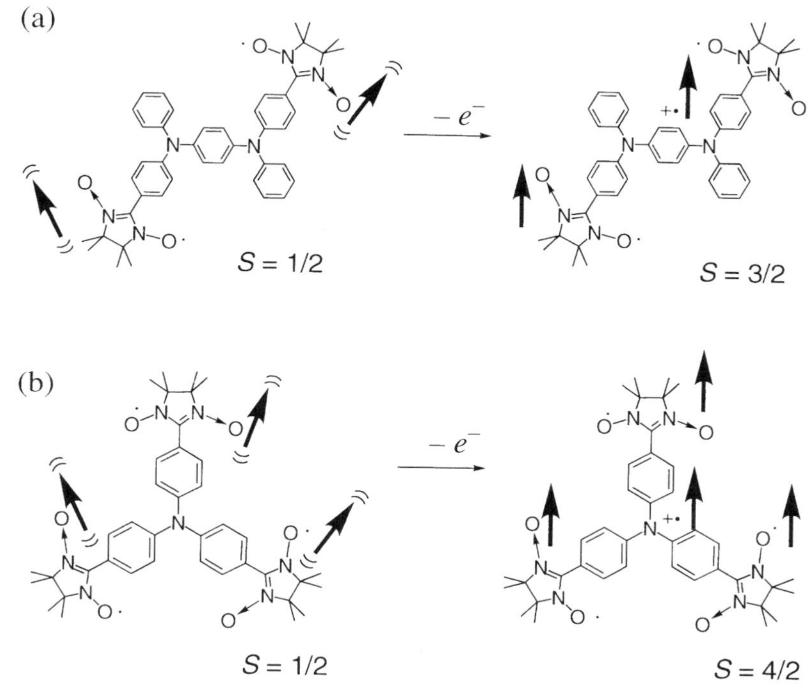

Figure 15 High spin cation polyradicals derived from (a) **PDNN₂** and (b) **TPANN₃**.

Oxidation of triphenylamine substituted by three NN groups (**TPANN₃**) by the addition of excess iodine to a THF solution afforded a multiplet ESR signal at 8 K. When the spectrum is assumed to be due to a quintet species, the $|D|$ value is calculated to be $0.0087\,\text{cm}^{-1}$ ($|E| = 0.0005\,\text{cm}^{-1}$). The temperature dependence of the high spin signal suggests that the quintet is the ground state of the oxidized species or it lies above the lower spin states by less than *ca.* $10\,\text{cal/mol}$ (Figure 15(b)).

The mechanism of the spin alignment is the same as that of **DMANN**. The high spin states of cation polyradicals may be derived from the positive exchange interaction between SOMO′, which spreads over the entire molecule, and SOMOs, which are localized on the radical units as depicted in Figure 16.

Although the spin systems are paramagnetic ($S = 1/2$) before oxidation, the one-electron oxidation at the central donor-core creates a ferromagnetic coupling between the spins on the stable radical sites to afford a high spin molecule. Such a novel spin system should be explored extensively in this research area.

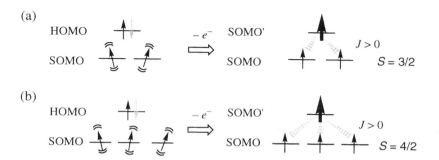

Figure 16 Possible mechanism of the spin alignment in polyradical cations: (a) **PDNN₂** and (b) **TPANN₃**. J denotes an intramolecular ferromagnetic interaction based on an positive exchange interaction between space-sharing MOs.

4. FERRIMAGNETIC SPIN SYSTEM COMPOSED OF "SPIN-POLARIZED" DONORS

Although ferrimagnetic spin systems of organic radicals containing transition metals are known,[18] purely organic ferrimagnets have not been constructed yet, except for a ferrimagnetic material composed of one-dimensionally ordered organic radicals with different spin multiplicities.[7] Mixed stacking of donors and acceptors is one of the basic assembling patterns of charge transfer complexes. When an organic acceptor and a *spin-polarized* donor are stacked alternately, ferrimagnetic interaction may be realized (Figure 7(b)).

The advantage of using *spin-polarized* donors for realizing such spin systems is as follows: the ferromagnetic intramolecular interaction (J_1 in Figure 7(b)) in the cation diradicals of *spin-polarized* donors is expected to be significantly large, related to the space-sharing character of singly occupied molecular orbitals. Namely, the high spin state of the cation diradical does not originate from the degeneracy of HOMOs depending upon the molecular symmetry,[19] but from the large exchange interaction between the two unpaired electrons which reside in two singly occupied molecular orbitals (SOMO and SOMO') of a space-sharing type. The triplet ground state of this species, therefore, does not suffer from a Jahn-Teller effect. Furthermore, the intermolecular antiferromagnetic interaction (J_2 in Figure 7(b)), which is based on the charge transfer, is expected to be much larger than that of through-space exchange interaction. Thus large intra- and intermolecular magnetic interactions may lead to a high T_c ferrimagnet.

A charge transfer complex of **APNN** and chloranil (**CA**) was obtained from benzene solution as blue plates.[5] An X-ray crystallographic analysis

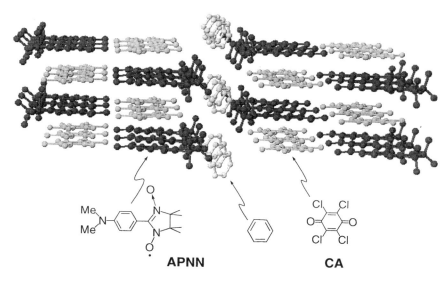

APNN   **CA**

Figure 17 Mixed stacks of **APNN** and **CA** in the charge transfer complex (**APNN · CA · C₆H₆**).

revealed that **APNN** and **CA** form a 1 : 1 complex with a mixed-stack columnar structure (Figure 17). A benzene molecule is incorporated within a cavity created by methyl groups of the nitronyl nitroxide moieties of the neighboring **APNN**s. The dimethylaminophenyl ring faces to the quinone ring of **CA** in parallel, the interplanar distance being 4.3 Å. This stacking indicates that HOMO of the donor and LUMO of the acceptor are overlapped exactly in phase. A broad peak, which is assignable to a charge transfer band, appeared at 900 nm in the absorption spectrum of the complex in a KBr disk.

Since **APNN** was proved to afford a ground state triplet cation diradical,[20] the complex may be considered to satisfy the requirement for the ferrimagnetic charge transfer complex. The complex, however, did not exhibit any ferrimagnetic interactions. The degree of charge transfer is not large enough to cause a sufficient contribution of the charge-transferred state into its electronic structure, due to the large difference in redox potentials of the donor ($E_{1/2}^{ox} = 0.64$ V) and the acceptor ($E_{1/2}^{red} = 0.11$ V). Another charge transfer complex was prepared, using **DMANN** ($E_{1/2}^{ox} = 0.53$ V) with a stronger donor ability and dichlorodicyanoquinone (**DDQ**) ($E_{1/2}^{red} = 0.59$ V). The complex turned out to be an ion radical salt, judged from the IR spectrum of the complex.[21] The complex, however, exhibited also a paramagnetic behavior with weak antiferromagnetic interaction

($\theta = -1$ K), presumably due to the large intermolecular antiferromagnetic interaction (J_2), compared with the intramolecular ferromagnetic interaction (J_1).

When the above experimental results were examined cautiously, the following requirement for a ferrimagnetic system may be pointed out. First, the degree of the charge transfer interaction should be large enough to afford a charge transfer complex with a charge transferred ground state. Second, the intramolecular ferromagnetic coupling (J_1) within the *spin-polarized* donor should be significantly larger than the intermolecular antiferromagnetic coupling (J_2). The appropriate combination of *spin-polarized* donors and acceptors should be perfected in future studies.

5. APPROACH TO ORGANIC FERROMAGNETIC METAL USING "SPIN-POLARIZED" DONORS

The *spin-polarized* donor developed here is, in fact, an important building block for constructing a conducting magnetic material as well (Figure 7(c)). When the *spin-polarized* donor is oxidized, the resulting unpaired electron is coupled ferromagnetically with the local spin on the radical site. If the columnar stacking of donors is realized, the conduction electrons, generated by hole doping, can migrate to align local spins on the radical sites (Figure 18). Although this approach was proposed by Yamaguchi,[11] it has not yet been realized.[22]

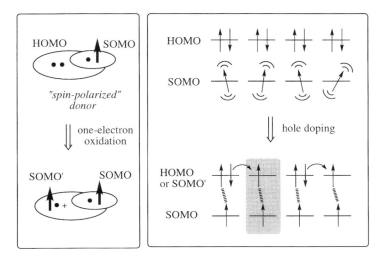

Figure 18   Spin alignment in organic ferromagnetic metal.

TTF-NN **TTF-PN** **EM-PN**

The electronic structure is closely related to the partially doped transition metal oxides such as $LaMnO_3$: manganese (III) in $LaMnO_3$ is of a high spin state.[23] Although $LaMnO_3$ is a Mott insulator and an antiferromagnet, it becomes conducting when it is doped with strontium, and the local spins line up through the conduction electrons (Figure 19). The spin alignment here is caused by the double exchange mechanism.

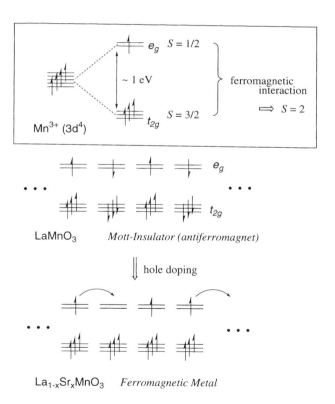

Figure 19 Spin alignment based on a double-exchange mechanism in $La_{1-x}Sr_x$-MnO_3.

In order to obtain a segregate columnar CT complex of a *spin-polarized* donor and also to increase the intermolecular overlap between radical donors, TTF derivatives carrying a NN group were prepared. A cation diradical derived from **TTF-NN** turned out to be a ground state singlet, although the thermally populated triplet signals were observed (Figure 20).[24] The antiferromagnetic interaction was estimated to be $J = -100$ K. The reason for the antiferromagnetic interaction may be explained as follows.

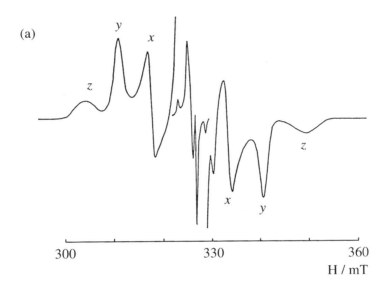

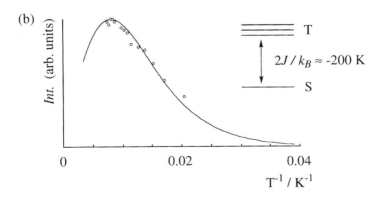

Figure 20 (a) Triplet ESR spectrum of **TTF-NN**$^{+\bullet}$ in MTHF glass at 110 K, (b) Temperature dependence of the triplet signal intensity of **TTF-NN**$^{+\bullet}$.

First, since the nitroxide group in NN is closely located to the sulfur atom of the TTF skeleton, the radical site is considered to be twisted from the donor plane. The twisting causes a significant decrease in the coefficients of HOMO at the radical site, breaking the condition of the space-sharing types of SOMOs. Second, a reverse CT may occur from the radical site to the cation radical of the donor site, the latter of which becomes an acceptor in the oxidized state.

To remove such an undesirable interaction, a *p*-phenylene group was inserted between the donor site and the radical site, in **TTF-PN**. Now the triplet was revealed to be the ground state for the cation diradical of **TTF-PN⁺·**. Based on these results, **EM-PN** was also prepared in order to enhance the kinetic stability and to increase the intermolecular interaction between cation diradicals; here an ethylendithio group was introduced at one end of the TTF moiety and a methylthio group was substituted at the olefinic carbon. The ground state of the cation diradical of **EM-PN** was proved to be triplet, and the oxidized species became reasonably stable.[25] The magnetic property of the charge transfer complexes of **EM-PN** will be elucidated in the near future.

CONCLUDING REMARKS

This chapter is devoted to introducing a basic study on the characteristic electronic structure of high-spin cation diradicals derived from *spin-polarized* donors. As described in Section 2, the cation diradical of **DMANN** turned out to exist as a ground state triplet species, and the electronic structure was revealed to be closely related to that of trimethylenemethane (**TMM**). While the unpaired electrons of **TMM** occupy two degenerated molecular orbitals, those of the cation diradical of **DMANN** occupy non-degenerated molecular orbitals; SOMO, which is localized on the stable radical site, and SOMO′, which spreads over the entire molecule. The extension of this spin system to the polyradical π-donors was also documented in Section 3. Namely, when a donor-core carrying stable radicals was singly oxidized, the polyradical donors turn out to afford high-spin cation polyradicals. If the redox process of these polyradicals is reversible, the interconversion becomes switchable between ferromagnetic and paramagnetic states. The molecular designing of *spin-polarized* donors described here may play an important role for actualizing organic ferrimagnets or organic ferromagnetic metals.

REFERENCES

1. For a recent overview, see: a) Sugawara, T. (1989). *J. Synth. Org. Chem., Jpn.*, **47**, 306; b) Kollmar, C. and Kahn, O. (1993). *Acc. Chem. Res.*, **26**, 259; c) Iwamura, H. (1990). *Adv. Phys. Org. Chem.*, **26**, 179; d) Miller, J.S. and Epstein, A.J. (1994). *Angew. Chem., Int. Ed. Engl.*, **33**, 385; e) Izuoka, A., Kumai, R. and Sugawara, T. (1995). *Adv. Mater.*, **7**, 672; f) Kahn, O. (1993). "*Molecular Magnetism*", VCH.

2. a) Itoh, K. (1967). *Chem. Phys. Lett.*, **1**, 235; b) Wasserman, E., Murray, R.W., Yanger, W.A., Trozzolo, A.M. and Smolinsky, G. (1967). *J. Am. Chem. Soc.*, **89**, 5076; c) Sugawara, T., Bandow, S., Kimura, K., Iwamura, H. and Itoh, K. (1986). *J. Am. Chem. Soc.*, **108**, 368.

3. a) McConnell, H.M. (1963). *J. Chem. Phys.*, **39**, 1910; b) Mukai, K., Nishiguchi, H. and Deguchi, Y. (1967). *J. Phys. Soc. Jpn.*, **23**, 125; c) Mukai, K. (1969). *Bull. Chem. Soc. Jpn.*, **42**, 40.

4. a) Izuoka, A., Murata, S., Sugawara, T. and Iwamura, H. (1985). *J. Am. Chem. Soc.*, **107**, 1786; b) Izuoka, A., Murata, S., Sugawara, T. and Iwamura, H. (1987). *J. Am. Chem. Soc.*, **109**, 2631.

5. Sugawara, T. and Izuoka, A. (1997). *Mol. Cryst. Liq. Cryst.*, **305**, 41.

6. a) Sugawara, T., Matsushita, M.M., Izuoka, A., Wada, N., Takeda, N. and Ishikawa, M. (1994). *J. Chem. Soc., Chem. Commun.*, 1723; b) Matsushita, M.M., Izuoka, A., Sugawara, T., Kobayashi, T., Wada, N., Takeda, N. and Ishikawa, M. (1997). *J. Am. Chem. Soc.*, **119**, 4369; c) Kinoshita, M., Turek, P., Tamura, M., Nozawa, K., Shiomi, D., Nakazawa, Y., Ishikawa, M., Takahashi, M., Awaga, K., Inabe, T. and Maruyama, Y. (1991). *Chem. Lett.*, 1225; d) Takahashi, M., Turek, P., Nakazawa, Y., Tamura, M., Nozawa, K., Shiomi, K., Ishikawa, M. and Kinoshita, M. (1991). *Phys. Rev. Lett.*, **67**, 746; e) Nakazawa, Y., Tamura, M., Shirakawa, N., Shiomi, D., Takahashi, M., Kinoshita, M. and Ishikawa, M. (1992). *Phys. Rev. B*, **46**, 8906; f) Chiarelli, R., Novak, M.A., Rassat, A. and Tholence, J.L. (1993). *Nature*, **363**, 147; g) Nogami, T., Ishida, T., Tsuboi, H., Yoshikawa, H., Yamamoto, H., Yasui, M., Iwasaki, F., Iwamura, H., Takeda, N. and Ishikawa, M. (1995). *Chem. Lett.*, 635; h) Cirujeda, J., Mas, M., Molins, E., Panthou, F.L., Laugier, J., Park, J.G., Paulsen, C., Rey, P., Rovira, C. and Veciana, J. (1995). *J. Chem. Soc., Chem. Commun.*, 709; i) Mukai, K., Konishi, K., Nedachi, K. and Takeda, K. (1995). *J. Magn. Magn. Mater.*, **140–144**, 1449; j) Caneschi, A., Ferraro, F., Gatteschi, D., Lirzin, A., Novak, M.A., Rentschler, E. and Sessoli, R. (1995). *Adv. Mater.*, **7**, 476.

7. a) Izuoka, A., Fukada, M., Sugawara, T., Sakai, M. and Bandow, S. (1992). *Chem. Lett.*, 1627; b) Izuoka, A., Fukada, M. and Sugawara, T. (1993). *Mol. Cryst. Liq. Cryst.*, **232**, 103; c) Izuoka, A., Fukada, M., Kumai, R., Itakura, M., Hikami, S. and Sugawara, T. (1994). *J. Am. Chem. Soc.*, **116**, 2609; d) Shiomi, D., Nishizawa, M., Sato, K., Takui, T., Itoh, K., Sakurai, H., Izuoka, A. and Sugawara, T. (1997). *J. Phys. Chem. B*, **101**, 3342.

8. a) Mukai, K., Konishi, K., Nedachi, K. and Takeda, K. (1996). *J. Phys. Chem.*, **100**, 9658; b) Takeda, K., Konishi, K., Nedachi, K. and Mukai, K. (1995). *Phys. Rev. Lett.*, **74**, 1673.

9. a) Mukai, K., Nuwa, M., Morishita, T., Muramatsu, T., Kobayashi, T.C. and Amaya, K. (1997). *Chem. Phys. Lett.*, **272**, 501; b) Kremer, R.K., Kanellakopulos, B., Bele, P., Brunner, H. and Neugebauer, F.A. (1994). *Chem. Phys. Lett.*, **230**, 255; c) Tomiyoshi, S., Yano, T., Azuma, N., Shoga, M., Yamada, K. and Yamauchi, J. (1994). *Phys. Rev. B*, **49**, 16031.

10. Awaga, K., Okuno, T., Yamaguchi, A., Hasegawa, M., Inabe, T., Maruyama, Y. and Wada, N. (1994). *Phys. Rev. B*, **49**, 3975.

11. a) Yamaguchi, K., Nishimoto, H., Fueno, T., Nogami, T. and Shirota, Y. (1990). *Chem. Phys. Lett.*, **166**, 408; b) Yamaguchi, K., Okumura, M., Fueno, T. and Nakasuji, K. (1991). *Synth. Metals*, **41**, 3631.

12. a) Sugawara, T. in "*Hyper-Structured Molecules I*", Sasabe H. (Ed.), Gordon & Breach, (1999). pp. 101–127; b) Dowd, P. (1972). *Acc. Chem. Res.*, **5**, 242; c) Borden, W.T. and Davidson, E.R. (1977). *J. Am. Chem. Soc.*, **99**, 4587; d) Borden, W.T., Iwamura, H. and Berson, J.A. (1994). *Acc. Chem. Res.*, **27**, 109.

13. Wenthold, P.G., Hu, J., Squires, R.R. and Lineberger, W.C. (1996). *J. Am. Chem. Soc.*, **118**, 475.

14. Sakurai, H., Kumai, R., Izuoka, A. and Sugawara, T. *Chem. Lett.*, **1996**, 879.

15. Dowd, P. (1966). *J. Am. Chem. Soc.*, **88**, 2587.

16. McConnell, H.M. (1961). *J. Chem. Phys.*, **35**, 1520.

17. Nakamura, T., Momose, T., Shida, T., Sato, K., Nakazawa, S., Kinoshita, T., Takui, T., Itoh, K., Okuno, T., Izuoka, A. and Sugawara, T. (1996). *J. Am. Chem. Soc.*, **118**, 8684.

18. a) Caneschi, A., Gatteschi, D., Sessoli, R. and Rey, P. (1989). *Acc. Chem. Res.*, **22**, 392; b) Caneschi, A., Gatteschi, D., Rey, P. and Sessoli, R. (1991). *Inorg. Chem.*, **30**, 3936; c) Kollmar, C., Couty, M. and Kahn, O. (1991). *J. Am. Chem. Soc.*, **113**, 7994; d) Stumpf, H.O., Ouahab, L., Pei, Y., Grandjean, D. and Kahn, O. (1993). *Science*, **261**, 447; e) Inoue, K. and Iwamura, H. (1994). *J. Chem. Soc., Chem. Commun.*, 2273.

19. a) LePage, T.J. and Breslow, R. (1987). *J. Am. Chem. Soc.*, **109**, 6412; b) Thomaides, J., Maslak P. and Breslow, R. (1988). *J. Am. Chem. Soc.*, **110**, 3970; c) Miller, J.S., Dixon, D.A., Calabrese, J.C., Vazquez, C., Krusic, P.J., Ward, M.D., Wasserman E. and Harlow, R.L. (1990). *J. Am. Chem. Soc.*, **112**, 381.

20. Kumai, R., Sakurai, H., Izuoka, A. and Sugawara, T. (1996). *Mol. Cryst. Liq. Cryst.*, **279**, 133.

21. Sakurai, H., Izuoka, A. and Sugawara, T. (1997). *Mol. Cryst. Liq. Cryst.*, **306**, 415.

22. a) Sugano, T., Fukasawa, T. and Kinoshita, M. (1991). *Synth. Metals*, **41**, 3281; b) Sugimoto, T., Yamaga, S., Nakai, M., Tsuji, M., Nakatsuji, H. and Hosoito, N. (1993). *Chem. Lett.*, 1817; c) Ishida, T., Tomioka, K., Nogami, T., Yamaguchi, K., Mori, W. and Shirota, Y. (1993). *Mol. Cryst. Liq. Cryst.*, **232**, 99; d) Nakatsuji, S., Hirai, A., Yamada, J., Suzuki, K., Enoki, T., Kinoshita, N. and Anzai, H. (1995). *Synth. Metals*, **71**, 1819; e) Nakatsuji, S., Akashi, N., Suzuki, K., Enoki, T., Kinoshita, N. and Anzai, H. (1995). *Mol. Cryst. Liq. Cryst.*, **268**, 153.

23. Kimura, T., Tomioka, Y., Kuwahara, H., Asamitsu, A., Tamura, M. and Tokura, Y. (1996). *Science*, **274**, 1698.

24. Kumai, R., Matsushita, M.M., Izuoka, A. and Sugawara, T. (1994). *J. Am. Chem. Soc.*, **116**, 4523.

25. Nakazaki, J., Matsushita, M.M., Izuoka, A. and Sugawara, T. (1999). *Tetrahedron letters*, **40**, 5027.

4. d-π MAGNETIC INTERACTIONS IN MCl$_4^{2-}$ (M = Mn(II) AND Co(II)) SALTS OF m- AND p-N-METHYLPYRIDINIUM NITRONYL NITROXIDES

AKIRA YAMAGUCHI and KUNIO AWAGA

Department of Basic Sciences, The University of Tokyo, 3-8-1 Komaba, Meguro-ku, Tokyo 153-8902, Japan

INTRODUCTION

Magnetic properties of a large number of transition metal complexes, organic radicals and metal-organic radical complexes have been studied so far, and the search for molecule-based magnetic materials has intensified in recent years.[1–4] In this field, a stable organic radical family, nitronyl nitroxide, is attracting much interest, because of potential ferromagnetic properties. Various nitronyl nitroxide derivatives have been found to exhibit ferromagnetic intermolecular interactions in their bulk crystals[5–21] since the discovery of the first pure organic ferromagnet, p-nitrophenyl nitronyl nitroxide.[22] Further, the nitronyl nitroxide has been also known to be a bidentate ligand for various transition and rare-earth metal ions. Ferromagnetic ground states have been observed in these complexes.[23–26]

Recently we have embarked upon the study of m- and p-N-alkylpyridinium nitronyl nitroxide (4,4,5,5-tetramethyl-2-(1-methyl-3 or 4-pyridinio)-3-oxide-4,5-dihydro-1H-1-imidazolyloxyl) radical cations. We have already reported magneto-structural correlations in the I⁻ and ClO$_4^-$ salts of m- and p-N-methylpyridinium nitronyl nitroxide (abbreviated as m- and p-MPYNN⁺, respectively).[13,14,27,28] m-MPYNN⁺I⁻ and m-MPYNN⁺ClO$_4^-$ are isostructural, in which m-MPYNN⁺ forms a bond-alternated hexagonal lattice involving both ferromagnetic and antiferromagnetic interactions.[13] The I⁻ and ClO$_4^-$ salts of p-MPYNN⁺ exhibit quite different crystal structures: the crystal of p-MPYNN⁺I⁻ consists of a radical dimer with an antiferromagnetic interaction, while, in the crystal of p-MPYNN⁺ClO$_4^-$, the p-MPYNN⁺ cation is isolated by the ClO$_4^-$ anion and shows Curie paramagnetism.[27] The crystal structures and magnetic properties of m- and p-MPYNN⁺ are expected to depend strongly on the counter anion. In this work we adopted the magnetic MCl$_4^{2-}$ (M = Mn²⁺($S = 5/2$) and Co²⁺($S = 3/2$)) anions as the counter of m- and p-MPYNN⁺. It is interesting to see the crystal structure caused by the insertion of MCl$_4^{2-}$, the d-π magnetic

59

interaction, and how the difference in the metal ion of MCl_4^{2-} affects the bulk magnetism. We describe the structural and magnetic properties of the four salts, $(m\text{-MPYNN}^+)_2$ $MnCl_4^{2-}$, $(m\text{-MPYNN}^+)_2CoCl_4^{2-}$, $(p\text{-MPYNN}^+)_2MnCl_4^{2-}$, and $(p\text{-MPYNN}^+)_2CoCl_4^{2-}$.

MATERIALS

The four salts were prepared as shown in Scheme 1. The iodide salts of $m\text{-MPYNN}^+$ and $p\text{-MPYNN}^+$ were prepared by the reported method.[8,29] The iodide ion was replaced by chloride in the following procedure. To an aqueous solution of p- or $m\text{-MPYNN}^+I^-$ (1 g, 2.7 mmol) was added an excess amount of powdered silver sulfate. During stirring of 1 hour, precipitation of silver iodide took place and the iodide ion was completely removed from the solution. To the filtrate including p- or $m\text{-MPYNN}^+$, and SO_4^{2-}, was added a solution of barium chloride. After 1 hour, precipitated barium sulfate was filtered off and the chloride salt was obtained by removing the solvent. A dry ethanol solution of m- or $p\text{-MPYNN}^+Cl^-$ (500 mg, 1.8 mmol) was added to a solution of $MnCl_2 \cdot 6H_2O$ (356 mg, 0.9 mmol) or $CoCl_2 \cdot 4H_2O$ (209 mg, 0.9 mmol), resulting in immediate

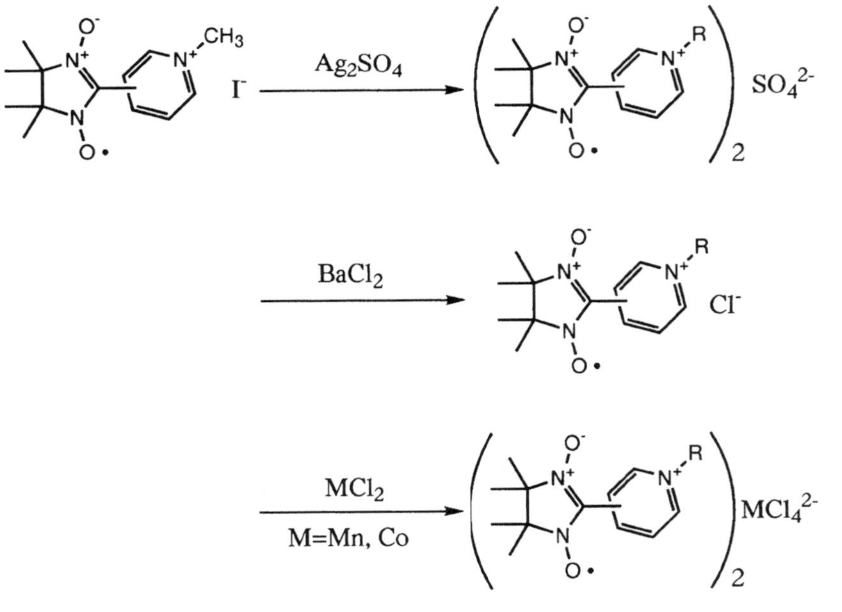

Scheme 1

precipitation of solid. The mixture was heated and a portion of hot ethanol was added until the precipitation dissolved. Slow evaporation of the solvent under a nitrogen atmosphere at room temperature led to crystallization of the MCl_4^{2-} salt. The $MnCl_4^{2-}$ salts were rather air-sensitive in contrast to the stable $CoCl_4^{2-}$ salts.

CRYSTAL STRUCTURES

The unit cell parameters of $(m\text{-MPYNN}^+)_2CoCl_4^{2-}$ and $(m\text{-MPYNN}^+)_2$ $MnCl_4^{2-}$ are compared in Table 1, where the ratios between the corresponding cell parameters are also listed. The two salts crystallize into a *monoclinic* system with the similar parameters. The cell volume of the $MnCl_4^{2-}$ salt is slightly larger than that of the $CoCl_4^{2-}$ salt, reflecting larger ion radius of Mn^{2+} than that of Co^{2+}. We have not carried out X-ray full structural analysis on the $MnCl_4^{2-}$ salt, but it is considered to be isostructural to the $CoCl_4^{2-}$ salt on which the structural refinement was done, because the X-ray diffractions of the $MnCl_4^{2-}$ salt in the range of $20 < 2\theta < 25°$ follow the same extinction rule as those of the $CoCl_4^{2-}$ salt do. Table 2 shows the crystal data of the two $p\text{-MPYNN}^+$ salts. They are also

Table 1 Crystal data for $(m\text{-MPYNN}^+)_2CoCl_4^{2-}$ and $(m\text{-MPYNN}^+)_2MnCl_4^{2-}$.

	$(m\text{-MPYNN}^+)_2CoCl_4^{2-}$	$(m\text{-MPYNN}^+)_2MnCl_4^{2-}$	Ratio
crystal system		monoclinic $P2_1/c$	
$a/\text{Å}$	16.607(3)	16.741(3)	1.008
$b/\text{Å}$	13.501(2)	13.538(3)	1.003
$c/\text{Å}$	14.370(2)	14.398(2)	1.002
$\beta/\text{deg.}$	90.43(2)	90.52(2)	1.001
$V/\text{Å}^3$	3221.8(9)	3262.9(9)	1.013

Table 2 Crystal data for $(p\text{-MPYNN}^+)_2CoCl_4^{2-}$ and $(p\text{-MPYNN}^+)_2MnCl_4^{2-}$.

	$(p\text{-MPYNN}^+)_2CoCl_4^{2-}$	$(p\text{-MPYNN}^+)_2MnCl_4^{2-}$	Ratio
		triclinic $P\bar{1}$-	
$a/\text{Å}$	18.010(2)	18.112(4)	1.006
$b/\text{Å}$	20.040(2)	20.147(4)	1.005
$c/\text{Å}$	9.918(1)	9.969(2)	1.005
$\alpha/\text{deg.}$	93.22(2)	93.19(2)	0.9997
$\beta/\text{deg.}$	99.17(1)	99.27(2)	1.001
$\gamma/\text{deg.}$	109.89(1)	110.11(2)	1.002
$V/\text{Å}^3$	3299.4(8)	3347(1)	1.014

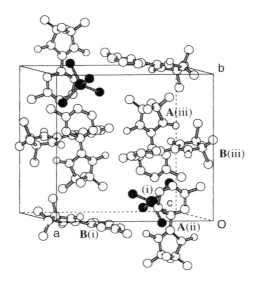

Figure 1   A view of the unit cell of $(m\text{-MPYNN}^+)_2\text{CoCl}_4^{2-}$. Symmetry operations: (i) x, y, z; (ii) $-x + 1$, $-y$, $-z$; (iii) $-x + 1$, $y + 1/2$, $-z + 1/2$.

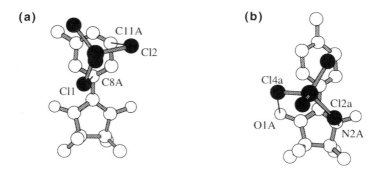

Figure 2   Intermolecular arrangements between the organic radicals and CoCl_4^{2-} in $(m\text{-MPYNN}^+)_2\text{CoCl}_4^{2-}$ (a) and $(p\text{-MPYNN}^+)_2\text{CoCl}_4^{2-}$ (b).

indicated to be isostructural, as the case of the $m\text{-MPYNN}^+$ salts. The crystal structures of the four salts depend little on MCl_4^{2-}.

Figure 1 shows a view of the unit cell of $(m\text{-MPYNN}^+)_2\text{CoCl}_4^{2-}$, where one CoCl_4^{2-} anion and two $m\text{-MPYNN}^+$ cations (molecules **A** and **B**) are crystallographically independent. The coordination of the cobalt(II) ion is slightly-distorted tetrahedral, and the CoCl_4^{2-} anion is surrounded by four $m\text{-MPYNN}^+$ cations, resulting in the isolation of CoCl_4^{2-}. Figure 2(a) shows

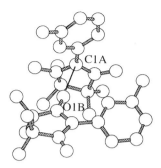

Figure 3 Intermolecular arrangement of the m-MPYNN$^+$ dimer in $(m\text{-MPYNN}^+)_2$ $CoCl_4^{2-}$.

a projection of $CoCl_4^{2-}$(i) onto the plane of the molecule **A**(ii), one of the four neighbors [symmetry operations: (i) x, y, z; (ii) −x + 1, −y, −z]. There are short intermolecular, interatomic distances between the chlorine atoms and the pyridinium rings: 3.61(2) Å for Cl1(i)···C8A(ii) and 3.61(2) Å for Cl2(i)···C11A(ii). They could be caused by electrostatic attraction forces. The arrangements between the $CoCl_4^{2-}$ anion and the other three m-MPYNN$^+$ cations are similar to that in Figure 2(a): the $CoCl_4^{2-}$ anion is located just on their pyridinium rings (not shown).

Furthermore, the crystal involves a short distance between **A** and **B** of m-MPYNN$^+$, whose arrangement is shown in Figure 3. The two molecular planes are oriented nearly perpendicular to each other. The shortest intermolecular, interatomic distance is 3.01(2) Å for C1A(i)···O1B(i). The magnetic interaction derived from this contact is discussed later.

Figure 4 shows a view of the unit cell of $(p\text{-MPYNN}^+)_2CoCl_4^{2-}$, where two $CoCl_4^{2-}$ anions (molecules **a** and **b**) and four p-MPYNN$^+$ cations (molecules **A–D**) are crystallographically independent. The bond lengths and angles are listed in Tables 1–2. The $CoCl_4^{2-}$ anions show slightly-distorted tetrahedral structures. Each $CoCl_4^{2-}$ is sandwiched by the two molecular planes of p-MPYNN$^+$. Figure 2(b) shows a projection of **a**(i) onto the molecular plane of **A**(i) [symmetry operation: (i) x, y, z]. The $CoCl_4^{2-}$ anion makes contacts with the NO groups, with shorter intermolecular, interatomic distances of 3.653(7) Å for Cl2a(i)···N2A(i) and 3.617(6) Å for Cl4a(i)···O1A(i). The molecule **B**(ii) is located at the other side of **A**(i) with respect to **a**(i), where there is also a distance between **B**(ii) and the NO group of **a**(i) [symmetry operation: (ii) −x + 1, −y + 1, −z + 1]. Further, **b**(i) of $CoCl_4^{2-}$ is sandwiched by the molecular planes of **C**(iii) and **D**(i) of p-MPYNN$^+$, having short distances to their NO groups [symmetry operation: (iii) x, y, z−1].

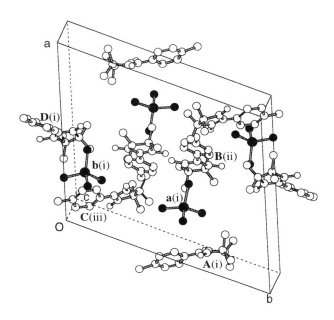

Figure 4   A view of the unit cell of $(p\text{-MPYNN}^+)_2\text{CoCl}_4^{2-}$. Symmetry operations: (i) x, y, z; (ii) $-x + 1$, $-y + 1$, $-z + 1$; (iii) x, y, z-1.

MAGNETIC PROPERTIES

The temperature dependence of the molar paramagnetic susceptibilities χ_p of $(m\text{-MPYNN}^+)_2\text{MnCl}_4^{2-}$ and $(m\text{-MPYNN}^+)_2\text{CoCl}_4^{2-}$ are shown in Figure 5(a), in the form of $\chi_p T$ vs. T. Each $\chi_p T$ value at room temperature is explained as that of the non-interacting magnetic moments on the organic and inorganic ions. In this figure, the $\chi_p T$ plots of the two salts appear parallel: $\chi_p T$ increases with decreasing temperature down to 10 K, indicating a ferromagnetic interaction, but it decreases below it. The two salts are concluded to involve similar magnetic interaction without depending on the metal ions. Figure 5(b) shows $\chi_p T$ of the two $p\text{-MPYNN}^+$ salts. They show opposite temperature dependence: $\chi_p T$ of $(p\text{-MPYNN}^+)_2\text{MnCl}_4^{2-}$ decreases with decreasing temperature, indicating an antiferromagnetic interaction, while that of $(p\text{-MPYNN}^+)_2\text{CoCl}_4^{2-}$ shows a ferromagnetic interaction. The magnetic interaction in the $p\text{-MPYNN}^+$ salts is seriously affected by the metal ions, whereas the magnetic properties of the $m\text{-MPYNN}^+$ salts depend little on MCl_4^{2-}.

In the low temperature region below 4 K, decreases of $\chi_p T$ are commonly observed in the ferromagnetic three salts. There are two possible reasons:

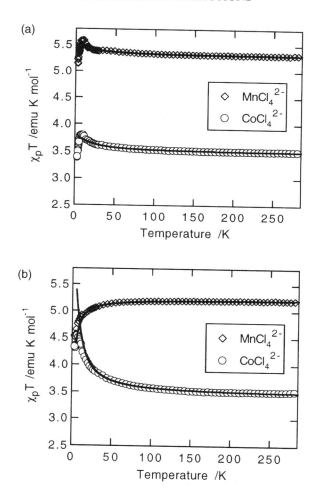

Figure 5 Temperature dependence of the paramagnetic susceptibilities of $(m\text{-MPYNN}^+)_2\text{MCl}_4^{2-}$ (a) and $(p\text{-MPYNN}^+)_2\text{MCl}_4^{2-}$ (b). The solid curves are the theoretical ones. See the text.

one is the zero field splitting on the metal ions, and the other is an anti-ferromagnetic coupling between the ferromagnetic units. At this stage, however, it is hard to identify the reason.

The structural difference between the corresponding CoCl_4^{2-} and MnCl_4^{2-} salts is thought to be so small that we discuss the magneto-structural correlations with the refined stuctures of the CoCl_4^{2-} salts. From the structural viewpoint, a crucial difference between the m- and p-MPYNN$^+$

salts is the position of the MCl_4^{2-} anion with respect to the molecular plane of the organic radicals. As shown in Figure 2, in the crystal of $(m\text{-MPYNN}^+)_2MCl_4^{2-}$, the arrangement is formed by a distance between MCl_4^{2-} and the pyridinium ring and there is no short distance between MCl_4^{2-} and the NO groups, while the crystals of $(p\text{-MPYNN}^+)_2MCl_4^{2-}$ involve the latter contact. The SOMO of the nitronyl nitroxide is localized on the two NO groups, making a node on the α-carbon bridging the two NO groups, and has little population in the aromatic substituent, i.e. the pyridinium ring in this case.[30] The arrangement in the m-MPYNN$^+$ salts, therefore, means little overlap of the SOMO of m-MPYNN$^+$ with the magnetic d orbitals of MCl_4^{2-}, suggesting the absence of the magnetic interaction between them. For this reason, the observed ferromagnetic interactions in the m-MPYNN$^+$ salts are expected to operate not between m-MPYNN$^+$ and MCl_4^{2-}, but between the organic radicals shown in Figure 3. This is consistent with the fact that the magnetic interaction in $(m\text{-MPYNN}^+)_2MCl_4^{2-}$ does not depend on MCl_4^{2-}. The intermolecular contact in the m-MPYNN$^+$ dimer is between the NO group and the α-carbon. This arrangement indicates a nearly orthogonal relation between the two SOMOs in spite of the short distance between them, which satisfy the requirement for the ferromagnetic intermolecular interaction.[30] In fact, similar arrangements have been observed in the ferromagnetic crystals of pyrimidinyl nitronyl nitroxide[8] and the lithium salt of p-benzoic acid nitronyl nitroxide.[12]

The temperature dependence of χ_p can be well interpreted as the sum of the ferromagnetic contribution from the m-MPYNN$^+$ dimer and Curie paramagnetic contribution from the MCl_4^{2-} anion, using

$$\chi_p = \frac{2N_A g_1^2 \mu_B^2}{k_B T\{3 + \exp(-2J/k_B T)\}} + \frac{N_A g_2^2 \mu_B^2}{3k_B T} S(S+1), \tag{1}$$

where J is the coupling constant in the ferromagnetic dimer, g_1 and g_2 are the g factors for the radical and the metal ion, respectively, N_A is the Avogadro constant, μ_B is the Bohr magneton, k_B is the Boltzmann constant and S is the spin quantum number of the metal ion. Using data above 30 K, the theoretical best fits are obtained with $J/k_B = 8.2\,\text{K}$, $g_1 = 2.00$ (fixed), and $g_2 = 2.04$ for the $MnCl_4^{2-}$ salt, and $J/k_B = 12.4\,\text{K}$, $g_1 = 2.00$ (fixed), and $g_2 = 2.41$ for the $CoCl_4^{2-}$ salt. The solid curves in Figure 5(a) are theoretical ones, and explain the observed behavior above 30 K.

In the crystals of $(p\text{-MPYNN}^+)_2MCl_4^{2-}$, on the other hand, there is a short distance between the MCl_4^{2-} anion and the NO group. This means an overlap between the magnetic d orbitals and the SOMO of the organic radical, which would be a dominant factor for realizing the bulk magnetism. Namely, the magnetic interaction between p-MPYNN$^+$ and MCl_4^{2-} is

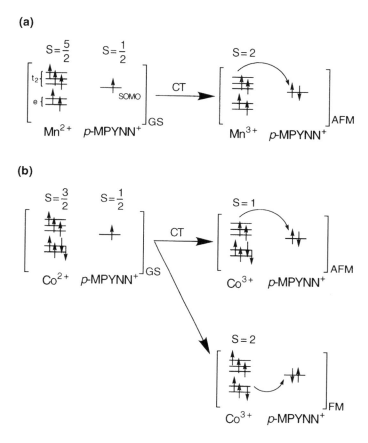

Figure 6 Electronic structures of the MCl_4^{2-} anions and the nitronyl nitroxide radical.

considered to change from antiferromagnetic to ferromagnetic by switching the metal ion in MCl_4^{2-} from Mn^{2+} to Co^{2+}. There is little difference in the crystal structure between $(p\text{-MPYNN}^+)_2CoCl_4^{2-}$ and $(p\text{-MPYNN}^+)_2$ $MnCl_4^{2-}$ and, therefore, the difference in the magnetic interaction would originate in the electronic structures of $MnCl_4^{2-}$ and $CoCl_4^{2-}$.

Figure 6 shows the electronic configurations of the divalent metal ions in a tetrahedral ligand field and the nitronyl nitroxide radical. The Mn^{2+} and Co^{2+} ions are in the $S = 5/2$ and $3/2$ ground states, respectively. Then we assume a charge transfer (CT) interaction from one of the $3d$ orbitals of the metal ion to the SOMO of the organic radical. A magnetic interaction between the $S = 5/2$ spin on Mn^{2+} and the $S = 1/2$ spin on $p\text{-MPYNN}^+$ results in an either $S = 3$ or 2 spin state. However the former one is forbidden in the CT interaction, because every frontier orbital is singly occupied in the ground state, so that the CT inevitably decrease the spin multiplicity down to $S = 2$, as shown in Figure 6(a). The resonance with the $S = 2$ CT excited state stabilizes an antiferromagnetic interaction. In the case of the $CoCl_4^{2-}$ salt, on the other hand, both ferromagnetic ($S = 2$) and antiferromagnetic ($S = 1$) excitations are possible (see Figure 6(b)). It is worth noting here that the metal ion in a tetrahedral ligand field always prefers a high spin ground state, because the electronic repulsion in the paired electrons is larger than the splitting energy of the tetrahedral ligand field. This means that the transfer which leaves higher spin multiplicity in the Co^{3+} (namely $S = 2$) is energetically advantageous. The admixture of the $S = 2$ CT excited state results in the stabilization of a ferromagnetic interaction between the $S = 3/2$ Co^{2+} ion and the $S = 1/2$ organic radical. For these, the difference in the magnetic interaction between the $MnCl_4^{2-}$ and $CoCl_4^{2-}$ salts can be qualitatively understood on the basis of the electronic structures of the metal ions.

Since MCl_4^{2-} is sandwiched by two p-MPYNN$^+$ cations in the crystals of (p-MPYNN$^+$)$_2$MCl$_4^{2-}$, we apply a three spin model, $H = -2J[S_1 \cdot S_2 + S_2 \cdot S_3]$, to the magnetic systems in the p-MPYNN$^+$ salts, where S_1 and S_3 are spin operators for the two p-MPYNN$^+$ cations, S_2 is that for the metal ion, and J is the coupling constant. For $S_1 = S_3 = 1/2$, the expression for χ_p is,[31]

$$
\chi_p = \frac{N_A g^2 \mu_B^2}{k_B T} \times \left\{ \frac{a_1(S_2) \exp(-(2S_2 + 2)J/k_B T)}{(2S_2 - 1) \exp(-(2S_2 + 2)J/k_B T)} \right.
$$
$$
\left. \frac{+a_2(S_2)(\exp(-2J/k_B T) + 1) + a_3(S_2) \exp(2S_2 J/k_B T)}{+(2S_2 + 1)(\exp(-2J/k_B T) + 1) + (2S_2 + 3) \exp(2S_2 J/k_B T)} \right\},
\tag{2}
$$

where g is the average g factor over the three spins, $S_2 = 3/2$, $a_1 = 1/2$, $a_2 = 5$, $a_3 = 35/2$ for (p-MPYNN$^+$)$_2$CoCl$_4^{2-}$, and $S_2 = 5/2$, $a_1 = 5$, $a_2 = 35/2$, $a_3 = 42$ for (p-MPYNN$^+$)$_2$MnCl$_4^{2-}$. Using data above 30 K, the theoretical best fits are obtained with $J/k_B = -0.72$ K and $g = 2.02$ for the MnCl$_4^{2-}$ salt, and $J/k_B = 3.6$ K and $g = 2.29$ for the CoCl$_4^{2-}$ salt. The theoretical curves in Figure 5(b) quantitatively explain the magnetic behavior of the p-MPYNN$^+$ salts above 30 K.

SUMMARY

We have studied the structural and magnetic properties of the four salts, (m-MPYNN$^+$)$_2$MnCl$_4^{2-}$, (m-MPYNN$^+$)$_2$CoCl$_4^{2-}$, (p-MPYNN$^+$)$_2$MnCl$_4^{2-}$, and (p-MPYNN$^+$)$_2$CoCl$_4^{2-}$. Their crystal structures are found to be governed only by the organic radical cations. A crucial difference in the mutual arrangement between the MCl_4^{2-} anion and the organic radical cation is that the MCl_4^{2-} is located just on the pyridinium ring of m-MPYNN$^+$ in the crystals of $(m$-MPYNN$^+)_2MCl_4^{2-}$, while the MCl_4^{2-} makes a contact with the NO groups of p-MPYNN$^+$ in $(p$-MPYNN$^+)_2$ MCl_4^{2-}. The two m-MPYNN$^+$ salts exhibit ferromagnetic behavior which is attributable to the m-MPYNN$^+$ dimer formed by the intermolecular distance between the NO group and the α-carbon. On the other hand, the two p-MPYNN$^+$ salts exhibit opposite magnetic behavior: $(p$-MPYNN$^+)_2$ $CoCl_4^{2-}$ and $(p$-MPYNN$^+)_2MnCl_4^{2-}$ are ferro- and antiferromagnetic, respectively. The opposite magnetic interactions in the p-MPYNN$^+$ salts are well understood in terms of the CT interaction between MCl_4^{2-} and p-MPYNN$^+$, in which the electronic structures of the Co^{2+} and Mn^{2+} ions are reflected.

REFERENCES

1. Miller J.S. and Epstein, A.J. (1994). *Angew. Chem., Int. Ed. Engl.*, **33**, 385.
2. *Molecule-Based Magnetic Materials: theory, techniques, and appplications*; eds., Turnbull, M.M., Sugimoto, T. and Thompson, L.K. (1996). American Chemical Society, Washington DC.
3. *Research Frontiers in Magnetochemistry*; ed., O'Connor, C.J. (1993). World Scientific Publishing, Singapore.
4. *Magnetism: A Supramolecular Function*; ed., Kahn, O. (1995). Kluwer Academic Publishers, Dordrecht.
5. Wang, H., Zhang, D., Wan, M. and Zhu, D. (1993). *Solid State Commun.*, **85**, 685.
6. Turek, P., Nozawa, K., Shiomi, D., Awaga, K., Inabe, T., Maruyama, Y. and Kinoshita, M. (1991). *Chem. Phys. Lett.*, **180**, 327.
7. Sugano, T., Tamura, M., Kinoshita, M., Sakai, Y. and Ohashi, Y. (1992). *Chem. Phys. Lett.*, **200**, 235.
8. Awaga, K., Inabe, T., Maruyama, Y., Nakamura, T. and Matsumoto, M. (1992). *Chem. Phys. Lett.*, **195**, 21.
9. Hernàndez, E., Mas, M., Molins, E., Rovira, C. and Veciana, J. (1993). *Angew. Chem., Int. Ed. Engl.*, **32**, 882.
10. Tamura, M., Shiomi, D., Hosokoshi, Y., Iwasawa, N., Nozawa, K., Kinoshita, M., Sawa, H. and Kato, R. (1993). *Mol. Cryst. Liq. Cryst.*, **232**, 45.
11. Panthou, F.L.d., Luneau, D., Laugier, J. and Rey, P. (1993). *J. Am. Chem. Soc.*, **115**, 9095.
12. Inoue K. and Iwamura, H. (1993). *Chem. Phys. Lett.*, **207**, 551.
13. Awaga, K., Okuno, T., Yamaguchi, A., Hasegawa, M., Inabe, T., Maruyama, Y. and Wada, N. (1994). *Phys. Rev. B*, **49**, 3975.
14. Awaga, K., Yamaguchi, A., Okuno, T., Inabe, T., Nakamura, T., Matsumoto, M. and Maruyama, Y. (1994). *J. Mater. Chem.*, **4**, 1377.
15. Matsushita, M.M., Izuoka, A., Sugawara, T., Kobayashi, T., Wada, N., Takeda, N. and Ishikawa, M. (1997). *J. Am. Chem. Soc.*, **119**, 4369.
16. Akita, T., Mazaki, Y., Kobayashi, K., Koga, N. and Iwamura, H. (1995). *J. Org. Chem.*, **60**, 2092.
17. Cirujeda, J., Ochando, L.E., Amigó, J.M., Rovira, C., Rius, J. and Veciana, J. (1995). *Angew. Chem., Int. Ed. Engl.*, **34**, 55.

18. Caneschi, A., Ferraro, F., Gatteschi, D., Lirzin, A.l., Novak, M.A., Rentschler E. and Sessoli, R. (1995). *Adv. Mater.*, **7**, 476.
19. Okuno, T., Otsuka T. and Awaga, K. (1995). *J. Chem. Soc., Chem. Commun.*, 827.
20. Imai, H., Inabe, T., Otsuka, T., Okuno, T. and Awaga, K. (1996). *Phys. Rev. B*, **54**, 6838.
21. Lang, A., Pei, Y., Ouahab, L. and Kahn, O. (1996). *Adv. Mater.*, **8**, 60.
22. Kinoshita, M., Turek, P., Tamura, M., Nozawa, K., Shiomi, D., Nakazawa, Y., Ishikawa, M., Takahashi, M., Awaga, K., Inabe, T. and Maruyama, Y. (1991). *Chem. Lett.*, 1225.
23. Gatteschi, D., Laugier, J., Rey, P. and Zanchini, C. (1987). *Inorg. Chem.*, **26**, 938.
24. Caneschi, A., Gatteschi, D., Rey, P. and Sessoli, R. (1988). *Inorg. Chem.*, **27**, 1756.
25. Caneschi, A., Gatteschi, D., Renard, J.P., Rey, P. and Sessoli, R. (1989). *Inorg, Chem.*, **28**, 2940.
26. Caneschi, A. and Gatteschi, D. (1991). *Prog. Inorg. Chem.*, **39**, 331. References are therein.
27. Yamaguchi, A., Awaga, K., Inabe, T., Nakamura, T., Matsumoto, M. and Maruyama, Y. (1993). *Chem. Lett.*, 1443.
28. Awaga, K., Inabe, T., Nakamura, T., Matsumoto, M. and Maruyama, Y. (1993). *Mol. Cryst. Liq. Cryst.*, **232**, 69.
29. Awaga, K., Inabe, T., Nagashima, U., Nakamura, T., Matsumoto, M., Kawabata, Y. and Maruyama, Y. (1991). *Chem. Lett.*, 1777.
30. Awaga, K., Inabe, T., Nagashima, U. and Maruyama, Y. (1989). *J. Chem. Soc., Chem. Commun.*, 1617.
31. Sekutowski, D., Jungst, R. and Stucky, G.D. (1978). *Inorg. Chem.*, **17**, 1848.

5. APPROACH TO HYPER-STRUCTURED LIGHT-TO-SPIN CONVERSION SYSTEMS

SHIGERU MURATA*

Department of Basic Sciences, Graduate School of Arts and Sciences, The University of Tokyo, 3-8-1 Komaba, Meguro-ku, Tokyo 153-8902, Japan

INTRODUCTION

Studies of high-spin organic molecules, which started from the discovery of the quintet dicarbene in 1967 (Itoh, 1967; Wasserman *et al.*, 1967), have been largely developed in recent years. Guiding principles which lead us to design an organic molecule with a high-spin multiplicity in its ground state have been theoretically established. According to them, a number of organic molecules with the high-spin quantum number have been prepared, and their magnetic properties have been examined.

Light can be conveniently employed for the preparation of high-spin organic molecules. In fact, the polycarbene with the spin quantum number of nine, which is the highest spin quantum number for organic molecules reported so far, was successfully prepared by the irradiation of the corresponding photolabile precursor (Nakamura *et al.*, 1993). It can be considered that, in these high-spin molecules, the photochemical energy is converted into the generation of magnetization. Though the efficiency of the energy conversion is highly important from the standpoint of energetics, no studies on the modification of the light employed to generate high-spin molecules were reported. Here we wish to propose a new hyper-structured molecular system, in which the photochemical energy can be effectively transferred into the magnetization of the high-spin molecule. This new hyper-structured light-to-spin conversion system is illustrated schematically in Figure 1. This system will operate in the three consecutive steps. At first, light is collected by antenna molecules. In the second step, photochemical energy is transferred into a precursor of the high-spin molecule. Finally, the species with a high-spin multiplicity is generated using the energy transferred from the antenna molecules. In order to construct this hyper-structured molecule, three units are required, that is, spin sources which can be induced photochemically, ferromagnetic couplers to construct a high-spin state, and antenna molecules. These three units should be optimized

---

* Tel.: 03-5454-6596, Fax: 03-5454-6998, E-mail: cmura@mail.ecc.u-tokyo.ac.jp

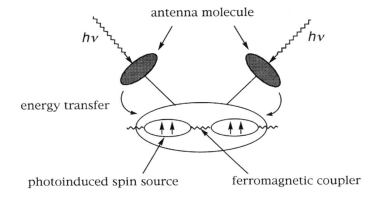

Figure 1 Hyper-structured light-to-spin conversion system.

independently, and then linked together to construct the hyper-structured system. In this chapter, the recent progresses in the researches concerning these three units are reviewed, which would be useful to design the hyper-structured light-to-spin conversion system.

SPIN SOURCES

Divalent carbon species (R_2C:) and univalent nitrogen species (RN:), which are called carbenes and nitrenes, respectively, have been established as photoinduced spin sources. In fact, studies of organic molecules with high-spin multiplicities have been largely developed by the use of carbenes as spin sources, as described in the following section. The advantages of these species as building blocks to construct high-spin molecules can be summarized in the following two points. First, carbenes and nitrenes can be readily generated by the irradiation of diazo compounds ($R_2C{=}N_2$) and azides ($RN{=}N_2$), respectively. In general, these precursors are thermally unstable compounds, while their thermal stability can be greatly enhanced by the phenyl substituents. Thus, diphenyl diazomethane ($Ph_2C{=}N_2$) and phenyl azide ($PhN{=}N_2$) are fairly stable at room temperature, and hence can be easily handled. Photolysis of these compounds with the light of a mercury lamp affords diphenylcarbene and phenylnitrene, respectively. Furthermore, the synthetic methods of these precursors are well-established; diphenyl diazomethane and phenyl azide can be readily prepared from benzophenone and aniline, respectively. Second, carbenes, as well as nitrenes, are one-centered diradical species, which have two unpaired electrons on the central atoms. In general, the two electrons of these species

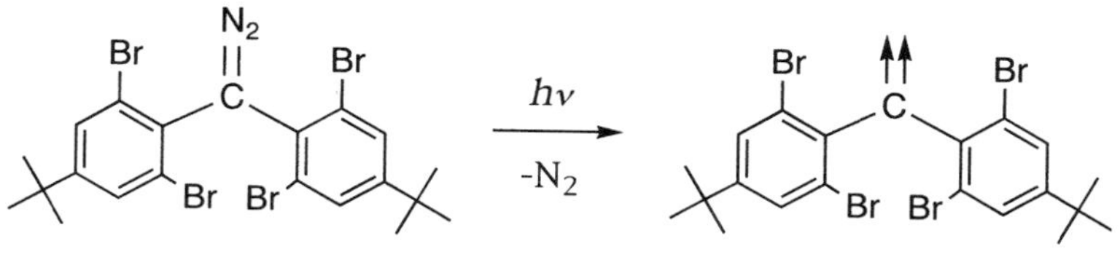

Figure 2 Generation of the superstabilized diphenylcarbene.

occupy two nearly degenerate molecular orbitals separately, so that these species have triplet ground states. The spin multiplicity of three in these species is favorable for building blocks to construct high-spin molecules, since the spin quantum number of polycarbenes and polynitrenes would be expected to double, compared with that of the corresponding polyradicals.

The disadvantage of these species is their thermal instability. Although diphenylcarbene and phenylnitrene are stable in organic matrices at cryogenic temperatures, these species disappear completely at around 100 K owing to their chemical reactions with surrounding molecules or dimerization. However, recently, the generation and characterization of superstabilized diphenylcarbene **1**, which is illustrated in Figure 2, have been reported by Tomioka and his co-workers (Tomioka *et al.*, 1996). The carbenic center of the carbene **1** is surrounded by four large bromine atoms, which prevent the chemical reactions with external reagents and dimerization of this carbene. They reported that the carbene **1** was stable up to $-110\,°C$ in matrices, and that, more surprisingly, it had a half-life period of 16 seconds at room temperature in solutions, which means that the carbene **1** is *ca.* 10^6 times more stable than unsubstituted diphenylcarbene. If superstabilized carbenes are employed as spin sources, the thermal stability of polycarbenes with high-spin multiplicities would be significantly increased, which enables to examine the magnetic properties of polycarbenes in the higher temperature ranges.

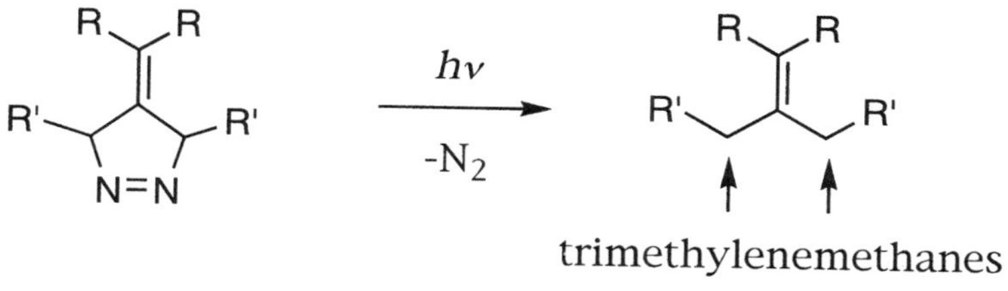

trimethylenemethanes

Figure 3 Generation of trimethylenemethane derivatives.

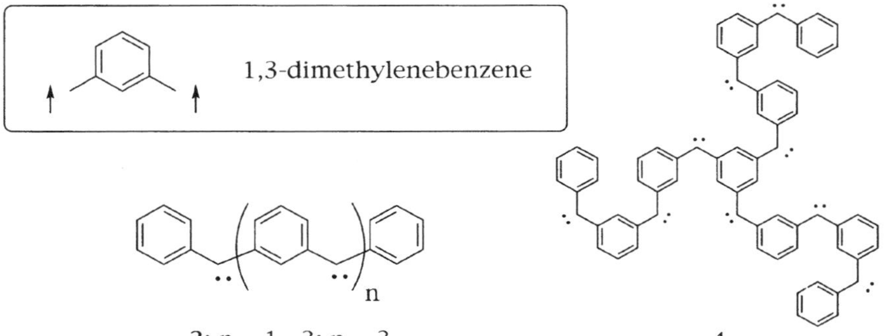

Figure 4 1,3-Dimethylenebenzene and polycarbenes with high-spin multiplicities.

Another useful spin sources which can be induced photochemically are trimethylenemethane derivatives. Trimethylenemethane is a π-diradical species with two degenerate molecular orbitals, the ground state of which has been experimentally determined to be triplet. Trimethylenemethanes can be generated by the irradiation of corresponding diazenes (Figure 3). Studies of high-spin molecules using these species as spin sources have been developed by Dougherty and his co-workers (Jacobs *et al.*, 1993).

FERROMAGNETIC COUPLERS

m-Bisphenylmethylene (**2**) is the first organic molecule established to have a quintet ground state (Itoh, 1967; Wasserman *et al.*, 1967). The dicarbene **2** consists of two phenylcarbene units attached to the meta positions of a benzene ring. It can be considered that 1,3-dimethylenebenzene, which is illustrated in Figure 4, couples two spin units ferromagnetically to construct a high-spin molecule. Theoretical background for the usefulness of 1,3-dimethylenebenzene as a ferromagnetic coupler has been discussed on the basis of molecular orbital theory and valence bond theory. According to the latter theory, the spin quantum number of the ground state of an alternant hydrocarbon, carbon atoms of which can be marked with a star alternatively, equals one-half of the difference between the numbers of starred and unstarred carbon atoms (Ovchinnikov, 1978). This rule, which would be called "a π-topology rule", is very useful for the prediction of the ground state spin multiplicity of an alternant hydrocarbon.

Since the discovery of the quintet dicarbene **2**, studies of high-spin organic molecules have been advanced by the use of 1,3-dimethylenebenzene as a ferromagnetic coupler. Thus, the tetracarbene **3** was strictly characterized to have a nonet ground state by means of ESR spectroscopy and the measurement of magnetic susceptibility (Sugawara *et al.*, 1986; Teki *et al.*, 1986). Recently, the nonacarbene **4** was synthesized and characterized to have the spin quantum number of nine in its ground state (Nakamura *et al.*, 1993).

Unfortunately, attempts to obtain polycarbenes with higher ground state spin multiplicities by the further extension of these molecules are unsuccessful at the present stage. Although 1,3-dimethylenebenzene is a useful ferromagnetic coupler, we should point out that there are some disadvantages to the high-spin systems based on this coupler. First, in these system, spin centers are located in the main framework of the molecule. Therefore, only when all sites are intact, the expected spin multiplicity would be realized. If one diazo group remains unconverted to the carbene or one carbene site is lost by chemical reactions or dimerization, the ferromagnetic coupling of the spins would be broken at this site, leading to the loss of the expected high-spin multiplicity. Second, there is no straightforward synthetic method leading to these systems. Therefore, synthesis becomes more and more difficult as the number of carbene sites is increased. In order to overcome these disadvantages, conjugated polymer chains, such as poly-(phenylacetylenes), poly(phenylenevinylenes) and poly(phenyldiacetylenes), have been noted as ferromagnetic couplers (Figure 5). In these systems, the spin sources reside on the side chain of a π-conjugated backbone. Although the high-spin polymers having carbenes or nitrenes as spin sources have not been obtained yet, the following two conclusions have been drawn from the studies using model compounds, which are illustrated in Figure 6 (Murata *et al.*, 1987; Murata and Iwamura, 1991). First, the interaction of two

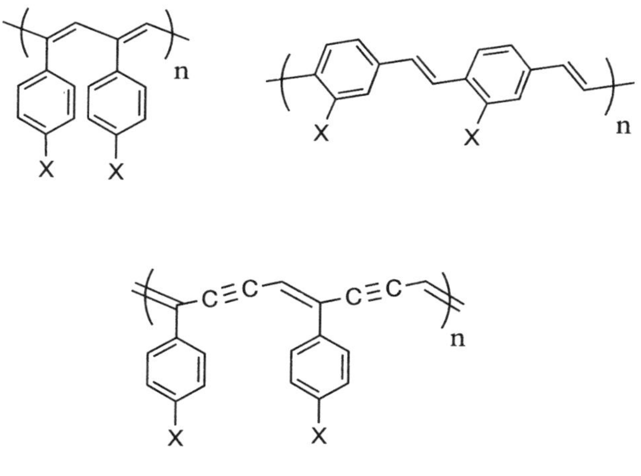

Figure 5 Conjugated polymer chains as ferromagnetic couplers (X denotes a spin source.)

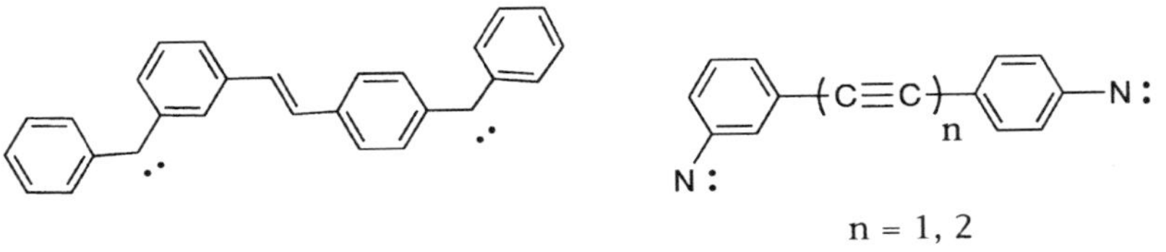

n = 1, 2

Figure 6   Model compounds of high-spin organic polymers.

remote triplet units connected through π-conjugated bonds is sufficiently strong to form the quintet state. Second, the ground state spin multiplicities in these systems are consistent with those predicted by "a π-topology rule" mentioned before. In order to realize the high-spin polymers with photo-induced spin sources, the design and synthesis of polymerizable monomer having precursors of the spin sources remain to be accomplished.

Dougherty and his co-workers evaluated experimentally that 1,3-cyclo-butane, as well as 1,3-cyclopentane, could be a ferromagnetic coupler (Jacobs *et al.*, 1993). These observations are supported by *ab initio* calcu-lations, while it is predicted that the ferromagnetic interaction of two spin units connected through these couplers is much smaller than the interaction through 1,3-dimethylenebenzene.

ANTENNA MOLECULES

Polycyclic aromatic compounds are good candidates for antenna molecules because of their large absorptivity in the longer wavelength ranges. How-ever, the question, which should be answered experimentally, has been raised whether the excited-state energy of an antenna molecule is sufficient to decompose a precursor of spin source. Figure 7 depicts a schematic energy diagram using pyrene and phenyl azide as an antenna molecule and a precursor of spin source, respectively. As shown in the figure, the energy transfer from excited pyrene to the electronically excited azide is an endo-thermic process, so that this energy transfer process would not be expected to proceed efficiently. However, there are two possibilities that the precursor of spin source is excited by the energy collected by the antenna molecule. First, if the electronically excited-state energy of the antenna molecule is effectively transferred to the vibrational mode of the azide moiety, the azide can be decomposed to give a nitrene, since the activation energy of the decomposition of phenyl azide is estimated to be only 35 kcal/mol. Second, if a single electron transfer occurs between the excited antenna molecule and the precursor of spin source, the precursor can be decomposed through the formation of its radical anion or radical cation.

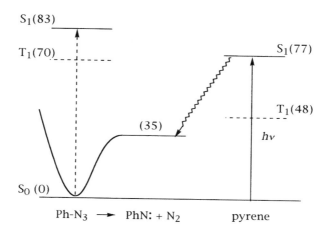

Figure 7 Energy diagram of an antenna molecule (pyrene) and a precursor of spin source (phenyl azide) (Energy levels are denoted in parentheses in kcal/mol.)

Unfortunately, very few studies have been reported on the photolysis of diazo compounds or azides sensitized by polycyclic aromatic compounds, though it is well-known that diazo compounds are decomposed to afford the corresponding triplet carbenes directly by the energy transfer from the triplet excited carbonyl compounds. In the 1960s, Lewis and Dalton reported that the photodecomposition of organic azides could be sensitized by polycyclic aromatic compounds, such as pyrene, triphenylene, and phenanthrene (Lewis and Dalton, 1969). They also pointed out that the sensitization occurred by a singlet energy transfer mechanism, while the energy transfer from sensitizers having lower excited-state energy than the azide proceeded more efficiently than expected for classical endothermic energy transfer. Recently, Murata and Tomioka reported the detailed mechanistic studies of pyrene-sensitized decomposition of p-butylphenyl azide in solutions (Murata *et al.*, 1995). Pyrene-sensitized photolysis of the azide in acetonitrile containing diethylamine gave the corresponding aniline, which could not be detected in the direct photolysis, together with the 3*H*-azepine derivative. The mechanism for the formation of these photoproducts is summarized in Figure 8, which involves a competitive quenching of singlet excited pyrene by energy transfer to the azide and electron transfer from diethylamine. These observations suggest that the singlet energy transfer from the excited polycyclic aromatic compounds to azides proceeds practically to afford nitrenes, though the mechanism of the energy transfer is controversial at present.

It would be reasonable to think that the efficiency of the energy transfer is increased with a decrease in the distance between an antenna molecule and a

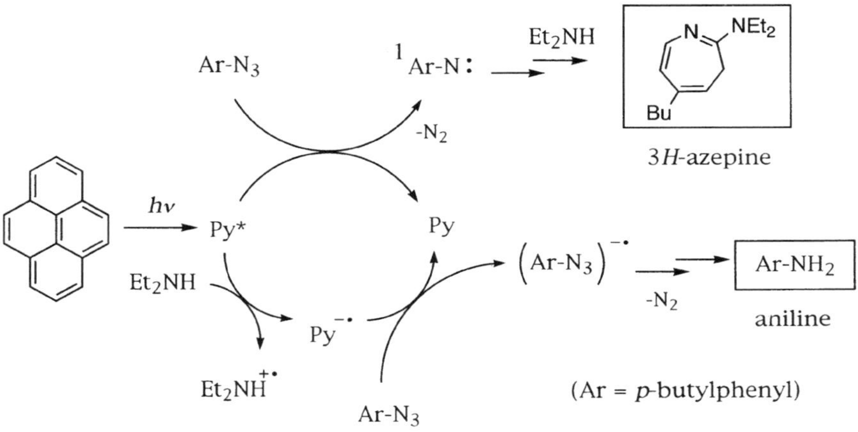

Figure 8 Mechanism of pyrene-sensitized decomposition of *p*-butylphenyl azide in the presence of diethylamine.

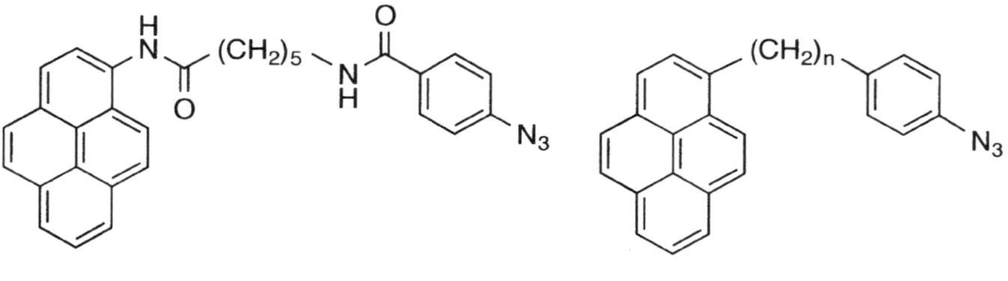

5   6: n = 1-5, 7, 13

Figure 9 Aryl azides covalently linked to an antenna molecule.

precursor of spin source. The photochemistry of aryl azides linked intra-molecularly to an antenna molecule **5**, which are illustrated in Figure 9, was first examined by Schuster and Buchardt (Shields *et al.*, 1988). They concluded that the intramolecular singlet energy transfer occurred in competition with the single electron transfer from an amidopyrene chromophore to the azide moiety, and suggested that the mechanism for energy transduction in the linked azides depended on the conformation of the molecule when it was excited. Recently, Murata and Tomioka prepared the aryl azides linked covalently to pyrene by a flexible methylene chain **6** (Figure 9), and examined their photodecomposition processes (Murata and Tomioka, 1995). On the basis of the emission spectra and the characterization of photoreaction products, the following two conclusions have been drawn. First, in the azide linked pyrene with a flexible methylene chain, the intramolecular singlet energy transfer from the excited antenna molecule to the azide moiety proceeds much more effectively than the intermolecular energy transfer. Second, the efficiency of the energy transfer is reduced with an increase in the number of methylene groups linking two moieties.

Although the observations reported so far have suggested that aryl azides can be decomposed to afford nitrenes using the energy collected by poly-cyclic aromatic compounds, several problems remain to be solved in order to design effective antenna molecules for light-to-spin conversion systems; (a) whether or not diazo compounds can be decomposed by the energy transfer from the excited polycyclic aromatic compounds to give carbenes, which are more convenient spin sources to construct high-spin molecules than nitrenes, (b) where is the lower limit of the excited-state energy of antenna molecules which are available to decompose a precursor of spin source, (c) whether or not the excitation of a precursor through the single electron transfer from or to the excited antenna molecule is available for the genera-tion of spins, (d) whether or not the excited-state energy of antenna mole-cules is sufficient to generate a number of spin units, though it is established that poly(diazo) groups decompose in rigid matrices at cryogenic tempera-ture to produce polycarbenes by the one-photon process, and (e) whether or not additional antenna molecules increase the efficiency of the light-to-spin conversion. A challenge to these problems is now ongoing.

Hyper-structured light-to-spin conversion system is of interest not only in the field of molecular-based magnetism, but of light-harvesting molecules. On the basis of the recent progresses in the researches on the three essential units of this system, that is, spin sources, ferromagnetic couplers, and antenna molecules, now is the time to approach the synthesis of this hyper-structured system.

REFERENCES

Itoh, K. (1967). *Chem. Phys. Lett.*, **1**, 235.

Jacobs, S.J., Shultz, D.A. Jain, R., Novak, J. and Dougherty, D.A. (1993). *J. Am. Chem. Soc.*, **115**, 1744.

Lewis, F.D. and Dalton, J.C. (1969). *J. Am. Chem. Soc.*, **91**, 5260.

Murata, S., Sugawara, T. and Iwamura, H. (1987). *J. Am. Chem. Soc.*, **109**, 1266.

Murata, S. and Iwamura, H. (1991). *J. Am. Chem. Soc.*, **113**, 5547.

Murata, S., Nakatsuji R. and Tomioka, H. (1995). *J. Chem. Soc., Perkin Trans.*, **2**, 793.

Murata, S. and Tomioka H. (1995). *Abstract of Papers*, Kyushu International Symposium on Physical Organic Chemistry, Fukuoka, 198.

Nakamura, N., Inoue, K. and Iwamura, H. (1993). *Angew. Chem., Int. Ed. Engl.*, **32**, 872.

Ovchinnikov, A.A. (1978). *Theoret. Chim. Acta* **47**, 297.

Shields, C.J., Falvey, D.E., Schuster, G.B., Buchardt O. and Nielsen, P.E. (1988). *J. Org. Chem.*, **53**, 3501.

Sugawara, T., Bandow, S., Kimura, K., Iwamura, H. and Itoh, K. (1986). *J. Am. Chem. Soc.*, **108**, 368.

Teki, Y., Takui, T., Itoh, K., Iwamura, H. and Kobayashi, K. (1986). *J. Am. Chem. Soc.*, **108**, 2147.

Tomioka, H., Hattori, M., Hirai, K. and Murata, S. (1996). *J. Am. Chem. Soc.*, **118**, 8723.

Wasserman, E., Murray, R.W., Yager, W.A., Trozzolo, A.M. and Smolinsky, G. (1967). *J. Am. Chem. Soc.*, **89**, 5076.

6. SYNTHESIS AND PHOTOPHYSICAL PROPERTIES OF HEPTANUCLEAR COMPLEXES WITH THREE DIMENSIONAL ROD-LIKE RIGIDITY

MASAHISA OSAWA, MIKIO HOSHINO and YASUO WAKATSUKI

*RIKEN (The Institute of Physical and Chemical Research),
Wako, 2-1 Hirosawa, Saitama 351-0198, Japan*

ABSTRACT

Successful preparation of novel dendrimer type complexes is described, where six peripheral polypyridine-Ru units are bound by acetylene and phenylene units to a central polypyridine-M (M = Ni, Cu, Ru) moiety. All three complexes have been found to have unique photophysical properties. While Cu is photo-inactive and works simply as a central binding spot, Ni placed in the center quenches MLCT state of the peripheral polypyridine-Ru units and implies potential utility of this type complexes for light harvesting devices. When Ru is placed in the center, interaction of the $(4d\pi)$ orbitals of the central Ru unit with high-lying ligand π^* orbitals has been noted which lead to delocalization of the π^* over whole dendric skeleton.

1. INTRODUCTION

Dendrimers which contain luminescent metal complexes have attracted considerable interest in recent years as novel photooptical materials potentially useful for light-harvesting devices.[1-3] Because of their outstanding excited states characteristics, Ru-polypyridine complexes are excellent candidates for photoactive building blocks. However, a majority of metallodendrimers having such species consist of flexible binding units like ether, amino, or methylene linkage[4-6] so that through bond interaction between metal moieties is not much expected. In the studies on one-dimensional rod-like complexes, importance of rigid bridges to link active components has been stressed because practical devices require vectorial energy or electron migration over long distances.[7-10] Following our previous finding that the acetylene bridged heterometallic complex $C_{p2}Ti(-\equiv-\equiv-Fc)_2$ ($C_p = \eta^5\text{-}C_5H_5$, Fc = Ferrocenyl), on oxidation of the Fc unit, exhibits remote charge transfer from the titanium bound carbon to Fc^+ through the $C\equiv C$ triple bonds,[11] we decided to construct three dimensional polynuclear ruthenium-polypyridine complexes with rigid and long rod-like linkers which connect the photoactive ruthenium units to a central metal.

2. RESULTS AND DISCUSSION

The mononuclear $\mathbf{1}\cdot 2PF_6$ was prepared by the reaction of (4-^{t}Bu-CC-terpy) ruthenium trichloride (terpy = 2, 2' : 6, 2''-terpyridine) with 4-$SiMe_3$-CC-terpy followed by addition of KPF_6. The bipy (bipy = 2,2'-bipyridine) linker was obtained by the Sonogashira type coupling reaction between 4, 4'-Br_2-bipy and two equiv H_2N-(C_6H_4)-CCH and then by changing the amino group to iodide with standard procedures. Another Sonogashira reaction of the resulting bipy derivative with $\mathbf{1}$ yielded $\mathbf{2}\cdot 4PF_6$ in 69% yield isolated as orange-red powder. IR(KBr, cm^{-1}): $\nu(C\equiv C)$, 2222, 2218. MALDI TOF-MS (m/Z): 2131 ($\mathbf{2}\cdot 3PF_6$), 1986 ($\mathbf{2}\cdot 2PF_6$), 1841 ($\mathbf{2}\cdot PF_6$), 1695 ($\mathbf{2}$). The presence of the ^{t}Bu-CC substituent at the 4-positions of the outer terpy is essential for good solubility of $\mathbf{2}$: without it, the dinuclear complex is insoluble in any solvent including DMF and DMSO. Finally three equiv $\mathbf{2}\cdot 4PF_6$ was reacted with MCl_2 (M = Ni, Cu) or $M(DMSO)_4Cl_2$ (M = Ru) followed by addition of KPF_6 and this convergent reaction was found to proceed fairly smoothly: 1 h reflux in acetonitrile in the cases of Ni and Cu and 3 h reflux in ethylene glycol in the case of Ru gave, after chromatographic purification, dark-red powder of the heptanuclear complexes $\mathbf{3}\cdot 14PF_6$ (M = Ni), $\mathbf{4}\cdot 14PF_6$ (M = Cu), and $\mathbf{5}\cdot 14PF_6$ (M = Ru), in excellent yields ($\mathbf{3}$ 95%, $\mathbf{4}$ 92%, $\mathbf{5}$ 70% based on $\mathbf{2}$). The new complexes were fully characterized by elemental analysis, FT-IR, and MALDI TOF mass spectrometry. MALDI TOF-MS (m/Z): 6450 ($\mathbf{3}\cdot 9PF_6$), 6308 ($\mathbf{3}\cdot 8PF_6$), 6164 ($\mathbf{3}\cdot 7PF_6$), 6017 ($\mathbf{3}\cdot 6PF_6$): 6457 ($\mathbf{4}\cdot 9PF_6$), 6317 ($\mathbf{4}\cdot 8PF_6$), 6171 ($\mathbf{4}\cdot 7PF_6$): 6785 ($\mathbf{5}\cdot 11PF_6$), 6641 ($\mathbf{5}\cdot 10PF_6$), 6495 ($\mathbf{5}\cdot 9PF_6$), 6354 ($\mathbf{5}\cdot 8PF_6$). Purity of complexes $\mathbf{3}$–$\mathbf{5}$ was further confirmed by ^{1}H-NMR spectra where any trace contamination of $\mathbf{2}$ was not detected. In complexes $\mathbf{3}$–$\mathbf{5}$, bonding of the central metal to the bipy domain of unit $\mathbf{2}$ is obvious: the three protons of the bipy rings in $\mathbf{2}\cdot 4PF_6$ are observed at δ 8.76 (H(6)), 8.56 (H(3)), and 7.59 (H(5)) in CD_3CN. In complex $\mathbf{3}\cdot 14PF_6$, these three protons are observed at δ 8.6, 7.6, and 7.4 as very broad resonance with no observable coupling due to the effect of paramagnetic Ni(II) core. Likewise, complex $\mathbf{4}\cdot 14PF_6$ exhibits extremely broad resonance at δ 8.2, 7.7, and 7.6. In complex $\mathbf{5}\cdot 14PF_6$ where the central metal is diamagnetic Ru(II), the bipy-proton resonance remain sharp at δ 8.72 (H(3)), 7.85 (H(6)), and 7.40 (H(5)). The structure of the products is thus assigned as illustrated in Scheme 1, the diameter of which should be *ca.* 60 Å.

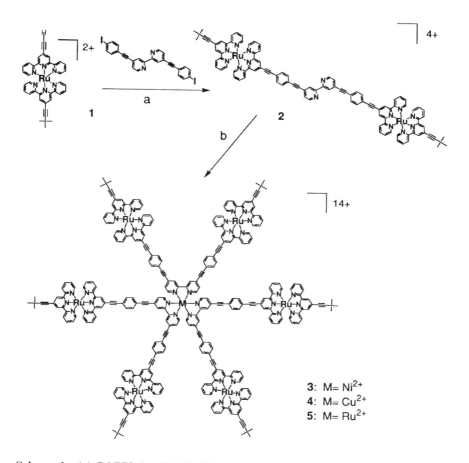

Scheme 1 (a) Pd(PPh₃)₄, CuI, Et₃N, in DMF. (b) NiCl₂6H₂O, CuCl₂2H₂O, or Ru(DMSO)₄Cl₂; CH₃CN for **3** and **4**, HOCH₂CH₂OH for **5**. The anion is PF₆.

The MLCT absorption bands and emission data of the new complexes **1**–**5** are summarized in Table 1. On going from **1** to **2**, both MLCT band and emission maximum shift to lower energy apparently as a result of extended conjugation along the longer π-system in **2**. Absorption, emission, and emission life time virtually do not change on assembling three units of **2** around M²⁺ indicating that the relative energy of the ground state and triplet excited state remains unchanged from **2** in the assembled complexes **3**, **4**, and **5**. However, though not given in the Table, a striking feature of the Ni-centered complex **3** was observed: its emission intensity is about one order of magnitude smaller than those of **4** and **5**. This low emission yield suggests that *ca.* 90% of photoenergy absorbed by the peripheral

Table 1 Absorption and emission data of complexes **1–5**, and [M(bipy)$_3$] · 2PF$_6$ (M = Ni, Cu).

	Absorption [a] λ, nm(ε, × 10^{4}M^{-1} cm^{-1})	Emission [b] λ, nm	Emission Life μs
1	483 (2.8)	645	4.0
2	497 (10)	680	10.0
3	498 (31)	685	7.1
4	497 (28)	680	8.3
5	499 (32)	690	4.0
[Ni(bipy)$_3$]$^{2+}$	525 (0.0012)[c]	—	—
	780 (0.0008)[c]	—	—
[Cu(bipy)$_3$]$^{2+}$	705 (0.0090)[c]	—	—

[a] MLCT absorption measured in CH$_3$CN at room temperature.
[b] Measured in *n*-C$_4$H$_9$CN at 150 K.
[c] Only d-d absorption was observed in CH$_3$CN. No absorption was observed at 355 nm where the laser excitation of **1–5** was performed for emission measurement.

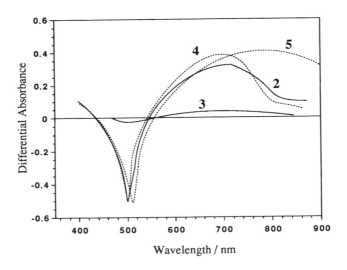

Figure 1 Differential absorption spectra recorded in *n*-C$_4$H$_9$CN at 150 K for the triplet excited states of complexes **2–5**.

Ru moieties is transmitted to the central Ni through the conjugated rod-like linker (Ru-Ni distance; *ca.* 20 Å) and ends up as radiationless decay.

The strong dependency of the excited-state properties on the central metal is also demonstrated in the triplet absorption spectra recorded after laser excitation at 355 nm (Figure 1). The very weak absorption of **3** centered at

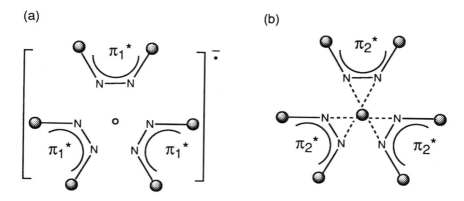

Figure 2 Schematic representation of two π^*-based orbitals of complex **5**. (a) The LUMO. (b) A higher lying vacant orbital. The triplet excitation is assumed to be (a) → (b) transition.

680 nm indicates the extremely low yield of the triplet state, in accord with the weak emission intensity mentioned above. The Cu-centered complex **4** shows a spectrum identical to that of **2** both in peak positions, centered at *ca*. 680 nm, and shape. The Ru-centered complex **5** is unique in that the corresponding absorption band is very broad and its center is shifted toward longer wavelength (780 nm).

According to the recent density functional calculations on $[Ru(bipy)_3]^{2+}$ by Daul *et al.* the lowest MLCT excitation is $(d-\pi : a_1) \rightarrow (\pi^* : a_2)$ transition, where the LUMO ($\pi^* : a_2$) is purely a ligand type while the higher vacant orbitals ($\pi^*:e$) contain samll but significant contribution of Ru(4d).[12] The broad band observed in the triplet absorption spectra of **5** (Figure 1), therefore, can be attributed to delocalization of the higher π^* orbitals of e type symmetry over the whole dendric skeleton via the central metal (Figure 2 (b)). In contrast the LUMO of a_2 type symmetry, which is populated on MLCT excitation of central Ru(bipy) moiety or more predominantly by excitation of one of the peripheral Ru(terpy) units, has an isolated structure of each component (Figure 2(a)). This simplified view is fully consistent with the identical emission maximum and band shape of complex **5** with those of unit **2** (Table 1). It has been manifested in the present study that tuning of photophysical properties of large-size metallodendrimers may be effectively attained by using rigid rod-like binding units and by changing the central metal.

REFERENCES

1. Balzani, V., Juris, A., Venturi, M., Campagna, S. and Serroni, S. (1996). *Chem. Rev.*, **96**, 759.
2. Bodige, S., Torres, A.S., Maloney, D.J., Tate, D., Kinsel, G.R., Walker, A.K. and MacDonnell, F.M. (1997). *J. Am. Chem. Soc.*, **119**, 10364 and references therein.
3. Lehn, J.-M. (1995). *Supramolecular Chemistry, Concepts and Perspective*, VCH, New York.
4. Constable, E.C., Harverson, P. and Ramsden, J.J. (1997). *Chem. Commun.*, 1683.
5. Constable, E.C. (1997). *Chem. Commun.*, 1073 and references therein.
6. Marvaud, V. and Astruc, D. (1997). *Chem. Commun.*, 773.
7. Vögtle, F., Frank, M., Nieger, M., Belser, P., Zelewsky, A., Balzani, V., Barigelletti, F., Cola, L.D. and Flamigni, L. (1993). *Angew. Chem. Int. Ed. Engl.*, **32**, 1643.
8. Harriman, A. and Ziessel, R. (1996). *Chem Commun.*, 1707.
9. Barigelletti, F., Flamigni, L., Collin, J.-P. and Sauvage, J.-P. (1997). *Chem. Commun.*, 333.
10. Barigelletti, F., Flamigni, L., Balzani, V., Collin, J.-P., Sauvage, J.-P., Sour, A., Constable, E.C. and Thompson, A.M.W.C. (1993). *J. C. S. Chem. Commun.*, 942.
11. Hayashi, Y., Osawa, M., Kobayashi, K. and Wakatsuki, Y. (1996). *Chem. Commun.*, 1617.
12. Daul, C., Baerends, E.J. and Vernooijis, P. (1994). *Inorg. Chem.*, **33**, 3538.

7. TUNING OF MOLECULAR TOPOLOGY IN CHROMOGENIC CYCLIC OLIGOMERS

TATSUO WADA[a,b,*], YADONG ZHANG[a], TAKASHI ISOSHIMA[a,b], YUJI KUBO[c] and HIROYUKI SASABE[a,d]

[a]Core Research for Evolutional Science and Technology (CREST), JST
[b]Supramolecular Science Laboratory, RIKEN (The Institute of Physical and Chemical Research), 2-1 Hirosawa, Wako, Saitama 351-0198, Japan
[c]Department of Applied Chemistry, Faculty of Engineering, Saitama University, 255 Shimo-Ohkubo, Urawa, Saitama 338-8570, Japan
[d]Department of Photonic Materials Science, Chitose Institute of Science and Technology, 758-65 Bibi, Chitose, Hokkaido 066-8655, Japan

ABSTRACT

Two types of chromogenic cyclic oligomers have been investigated in terms of the fine adjustment or "tuning" of their molecular topology, specifically for the case of carbazole cyclic oligomers and indoaniline-derived calix[4]-arenes. In the Knoevenagel polycondensation diformyl compounds reacted with 9-alkyl-3,6-bis(cyanoacetoxymethyl)carbazole to yield a main-chain polymer or cyclic oligomer. It is noteworthy that the main products can be controlled by the condensation conditions and the ring size of cyclic oligomers can be determined by the bis(cyanoacetate) structures. Efficient energy transfer was observed in cyclic oligomers with alternating units of acceptor-substituted carbazole moieties and electron donative ones connected through 3,6-linkages. The topological shapes of indoaniline-derived calix[4]arenes were studied by hyper-Rayleigh scattering which provides information on the hyperpolarizability of the molecule. The two indoaniline moieties in calix[4]arene derivatives were pre-aligned and net molecular hyperpolarizability was enhanced. Besides dendritic oligomers, cyclic oligomers can be used as a molecular platform which allows the molecular level tuning of shape, size and topology for superior opto-electronic functions.

INTRODUCTION

In the field of arts and designs, it is well known that "form follows functions". This philosophy also holds true in the development of functional

* Tel.: +81-48-467-9378, Fax: +81-48-462-4647, E-mail: tatsuow@postman.riken.go.jp

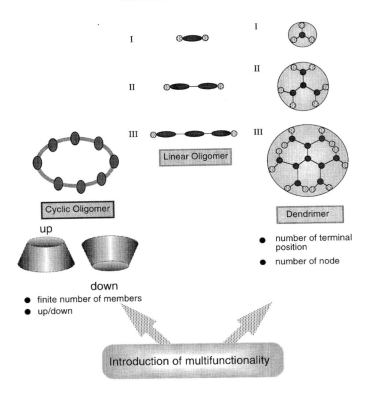

Figure 1 Generation of dendrimers, linear and cyclic oligomers.

materials. Previously we have described our research approach to organic quantum devices using hyper-structured molecules (HSMs) based on carbazole multifunctional oligomers such as carbazole trimers and dendrimers.[1] In this study we discuss the fine adjustment or "tuning" of the molecular topology of HSMs, especially for the case of the cyclic oligomers.

One of the most interesting and important features in the dendritic molecules is their divergent structures, that is, the number of nodes increases depending on the generation as shown in Figure 1. In contrast to linear chain oligomers, the large number of terminal positions can be obtained. By introducing electroactive groups into the terminal positions of dendrimers, we can modify the total features of these dendrimers chemically and physically. While there is a structural freedom in linear chain oligomers, the molecular shape of dendrimers can be controlled by the structure of the spacer units and the generation. Using rigid spacer groups, we can obtain stretched linear oligomers, however, these rod-like molecules show poor processability. On the other hand, it was reported that higher generation dendrimers are quite soluble in common organic solvents.[2] In the case of cyclic oligomers, opto-electronic functional groups can be introduced as a member of a ring. Interaction through space and through bonds among these functional groups depends on the ring size and the sequence of ring members. In addition to ring size, the topological shape of a ring can be controlled using specific cyclic compounds such as chromogenic calixarenes. Besides the feasibility of chemical modification, interesting physical properties can be expected in these oligomeric systems.

TOPOLOGY AND FUNCTIONS

In terms of opto-electronic functions, the well studied functional cyclic compounds are porphyrins and related compounds. These functional chromophores consist of tetra pyrrole derivatives. Their unique opto-electronic properties originate not in the nature of the components but in the conjugated two-dimensional π-electrons. On the other hand, cyclic oligomers in which ring members are connected with saturated bonds have no two-dimensional conjugation. Unlike porphyrins, one cannot expect opto-electronic properties of the conjugated two-dimensional π-electrons for these cyclic oligomers. However, interesting host-guest interactions are observed in these cyclic oligomers such as cyclodextrin and crown ethers. Research on crown ethers is not limited to preparative organic chemistry alone but is an interdisciplinary field. In these compounds, their topology is highly correlated to their functions.[3] From a topological point of view, neutral ligands of crown ethers can be classified into three groups: open-chain compounds, known as *podands*, monocyclic systems, called *coronands*, and oligocyclic (spherical) ligands with the name *cryptands*. The ligand topology is fundamentally responsible for the complexation properties of the crown ethers. In monocyclic systems focused in this study, there are four different topology: (a) partially aligned, (b) cone, (c) Möbius strip and (d) flat disk as illustrated in Figure 2. We introduced opto-electronic active sites into a ring member. Two types of molecular systems were studied as HSMs: carbazole cyclic oligomers and indoaniline-derived calix[4]arene as shown in Figure 3. Although there is no two-dimensional conjugation along the cyclic path, we can expect unique opto-electronic properties which is sensitive to the topology of the ring.

CYCLIC OLIGOMERS

Recently we have developed the efficient synthesis of a novel carbazole cyclic oligomer and a main-chain polymer via the Knoevenagel condensation

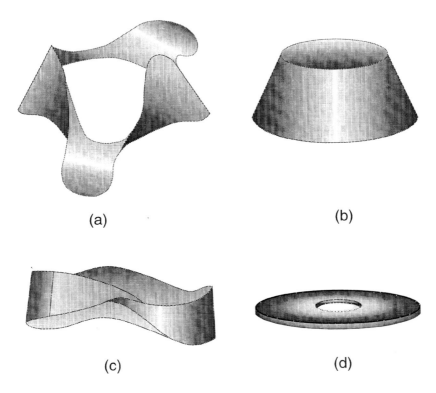

(a)   (b)

(c)   (d)

Figure 2   Topology of cyclic compounds.

as shown in Scheme 1.[4] The carbazole cyclic tetramer could be obtained in a high yield by a one-stage Knoevenagel condensation of 3,6-diformyl-9-heptylcarbazole and 3,6-bis(cyanoacetoxymethyl)-9-heptyl-carbazole in tetrahydrofuran (THF) without the use of the high dilution principle. In the polycondensation of diformyl and cyanoacetate, there are several interesting issues: the main products could be obtained in a polymer or a cyclic oligomer by controlling the reaction conditions. The cyclic oligomer was always obtained in THF reaction solution as a major product in a high yield. Size-exclusive chromatography shows a sharp peak, which means that the cyclic oligomer has monodispersion of molecular weight and only one kind of cyclic oligomer (tetramer in this case) was obtained from this condensation reaction (Figure 4). The carbazole main-chain polymer with very large molecular weight also could be obtained in a high yield by a two-stage Knoevenagel polycondensation. The polycondensation in THF followed by solid state polycondensation after removal of THF yielded the polymer as a main product.

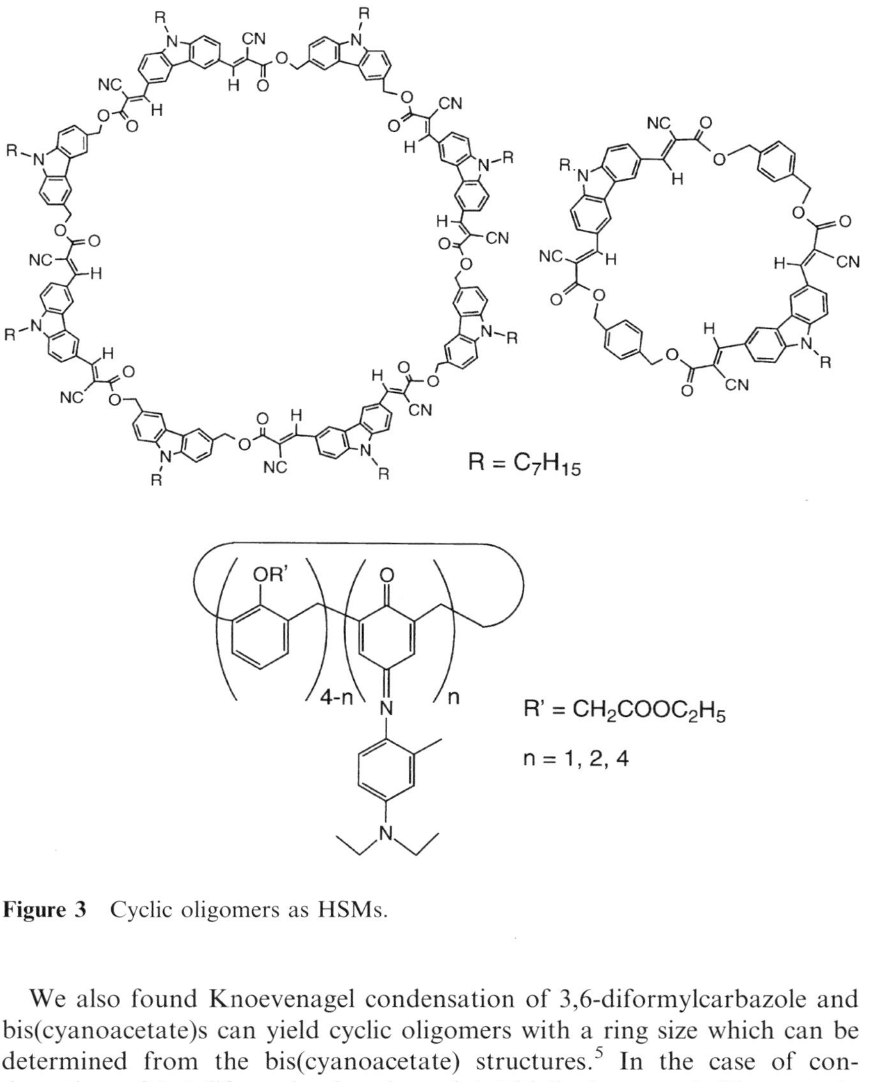

Figure 3 Cyclic oligomers as HSMs.

We also found Knoevenagel condensation of 3,6-diformylcarbazole and bis(cyanoacetate)s can yield cyclic oligomers with a ring size which can be determined from the bis(cyanoacetate) structures.[5] In the case of condensation of 3,6-diformylcarbazole and 1,4-bis(hydroxymethyl)benzene in THF solution, a yellow cyclic dimer was obtained as a major product in 78% yield (as shown in Scheme 1). It was also found that the carbazole main-chain polymer could be obtained in 94% yield by a two-stage Knoevenagel polycondensation.[6] The monomers and linear oligomers yielded by the polycondensation are soluble in THF. This condensation is a reversible reaction in the presence of a base, and hence this is the reason why we can not obtain carbazole main-chain polymer from the solution condensation.

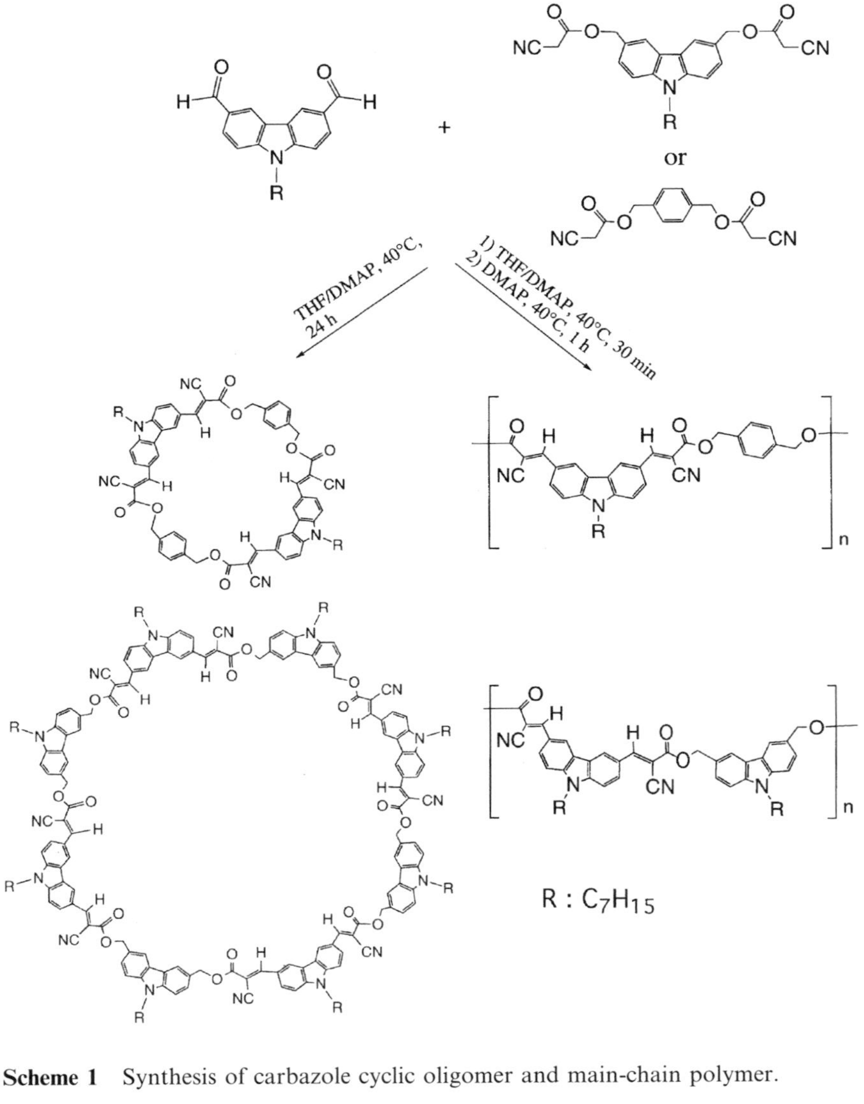

Scheme 1 Synthesis of carbazole cyclic oligomer and main-chain polymer.

On the other hand, the cyclic oligomer is insoluble in THF. Once the cyclic oligomer was produced, the cyclic oligomer precipitated from the reaction solution. Therefore, the chemical equilibrium must be favorable for the formation of the cyclic oligomer in THF solution. This is the reason why the cyclic oligomer can be obtained in a high yield. In general, the production of high molecular weight polymers is often accompanied by the formation of

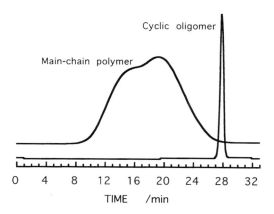

Figure 4 Size-exclusive chromatography.

cyclic oligomers with various ring sizes in polycondensation reaction.[7] However, in our case, only one size of the cyclic oligomer was obtained.

The cyclic oligomer has alternating units of acceptor-substituted and electron donative carbazole moieties connected through 3,6-linkages. The fluorescence peaks of monomeric carbazole were generally observed at 355 and 370 nm assigned to the (0,0) transition and its vibronic band, respectively.[8] After introduction of acceptor groups such as CH= C(CN)COOR into 3 and/or 6 positions of carbazole ring, the intracharge-transfer absorption band was observed around 420 nm depending on the acceptor groups. Fluorescence peak of acceptor-substituted carbazole was observed at ~ 460 nm excited at 300–420 nm. Here we can expect an efficient energy transfer from the excited state of carbazole moieties to the acceptor-substituted carbazole moieties in a cyclic oligomer. Figure 6 shows the absorption, fluorescence and fluorescence excitation spectra monitored at the registration wavelength of 460 nm for carbazole tetramer. While there is an optical window above 400 nm for carbazole moieties in the tetramer, both carbazole and the acceptor-substituted carbazole moieties can be excited at 300–350 nm. Fluorescence from the carbazole moieties excited at 300–350 nm was quenched and only the fluorescence for acceptor-substituted carbazole moieties was observed. These acceptor-substituted carbazole moieties behaved as an energy acceptor as shown in Figure 5. Therefore relative excitation yields of cyclic oligomers increased at 300–350 nm in comparison to those of a carbazole monomer. We can also expect a different energy transfer process along the cyclic path in cyclic oligomers in comparison with one-dimensional random walk in the polymer which has alternating units of acceptor-substituted and electron donative carbazole moieties. Moreover this type of acceptor-substituted carbazole moieties

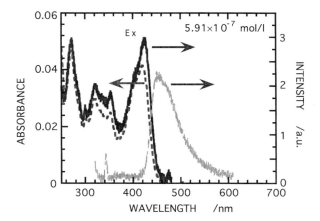

Figure 5   Absorption, fluorescence and fluorescence excitation (Ex) spectra for the carbazole cyclic oligomer solution.

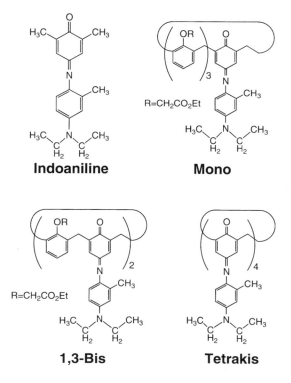

Figure 6   Chemical structures of indoaniline and indoaniline-derived calix[4]arenes.

displayed electroluminescent properties based on their elecron-transporting and fluorescent properties.[9]

INDOANILINE-DERIVED CALIX[4]ARENES

Calix[4]arenes are molecules which receive increasing attention in the field of supramolecular chemistry.[10] As a chromogenic calix[4]arene, we describe the indoaniline-derived calix[4]arene with different numbers of indoaniline moiety. Originally this chromogenic calix[4]arene was developed by Kubo *et al.* as an artificial receptor: host-guest interaction induced the change of absorption for sensing by a chromophore group due to a perturbation of the electronic state.[11] Synthesis is quite straightforward except for separation of number of isomers. Chromogenic calix[4]arenes with a different number of indoaniline moiety were synthesized from the condensation between calix[4]arene and 4-(diethyl-amino)-2-methylaniline hydrochloride under alkaline conditions in the presence of $K_3[Fe(CN)_6]$ as an oxidizing agent. Calix[4]arene with two indoaniline groups in Figure 6 has four different conformers: cone, partial cone and their alternates. In order to estimate the conformation of calix[4]arenes, we determined the hyperpolarizabilities of indoaniline-derived calix[4]arenes using the hyper-Rayleigh scattering (HRS) technique[12] because the second-order nonlinear optical responses are very sensitive to the symmetry of molecules. Hyperpolarizability (β) is related to the difference between the ground state and excited state dipole moments, and oscillator strength. To obtain a compound with a large β value, we must optimize each parameter for β. Indoaniline compound has a larger β value than *para*-nitro aminobenzene and donor-acceptor stilbene due to the contribution of mixed benzene and quinoid structures in the ground state.[13] Since indoaniline derivatives have red-shifted absorption band (λ max : 610 nm), HRS signal (532 nm) using a YAG laser as a fundamental light located in the absorption band. Therefore one may argue the multi-photon emission from indoaniline derivatives besides HRS signals. A time-resolved subpicosecond HRS study, however, showed that we could distinguish HRS signals from the multi-photon fluorescence.[14] Figure 7 shows the typical emission spectrum of indoaniline solution pumped using a subpicosecond pulse. Besides HRS signal, there was an emission at around 620 nm. Table 1 summarized β values of various compounds determined by the HRS technique including those of *para*-nitro aminobenzene (PNA) and *para*-nitro dimethyl aminobenzene (DMA) for comparison. The β value of 870×10^{-30} esu for indoaniline was obtained. In the case of mono-substituted calix[4]arene, the similar value was obtained. On the other hand, those of 1,3-bis- and tetrakis(indoaniline)-derived calix[4]arenes were 1300×10^{-30} esu. Enhancement factor was less than 2, but at least two

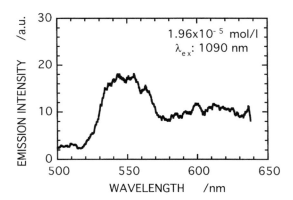

Figure 7 Emission spectrum of indoaniline solution pumped by using a subpicosecond pulse (excitation wavelength λ_{ex} : 1090 nm).

Table 1 Hyperpolarizabilities determined by an HRS technique.

Compound	$\beta/10^{-30}$ esu
Indoaniline	870
Mono	860
1,3-Bis	1,300
Tetrakis	1,300
PNA	29
DMA	63

indoaniline sites in these calix[4]arenes were pre-aligned in the solution. The topology of 1,3-bis- and tetrakis(indoaniline)-derived calix[4]arenes could be type (a) in Figure 2. Using calix[4]arene as a molecular platform, we can also control the topological shape of cyclic oligomers and apply these systems to second-order nonlinear optical molecules.

CONCLUSION

We have successfully developed cyclic oligomers as hyper-structured molecules and demonstrated the capability to control the ring size and topology. Efficient energy transfer and enhanced second-order nonlinear optical responses were obtained in these cyclic oligomers.

REFERENCES

1. Wada, T., Zhang, Y. and Sasabe, H. chapter 1 in this book.
2. Tomalia, D.A., Naylor, A.M. and Goddard III, W.A. (1990). *Angew. Chem. Int. Ed. Engl.*, **29**, 138.
3. Vögtle, F. (1991). *Supramolecular Chemistry*, John Wiley & Sons.
4. Zhang, Y., Wada, T. and Sasabe, H. (1996). *Chem. Commun.*, 621.
5. Wada, T., Zhang, Y., Aoyama, T. and Sasabe, H. (1997). *Proc. Japan Acad.*, **73**, 165.
6. Zhang, Y., Wang, L., Wada, T. and Sasabe, H. (1996). *Macromolecules*, **29**, 1569.
7. Montaudo, G., Scamporrrino, E.J., Puglisi, C. and Vitalini, D. (1987). *Polym. Sci., Part A: Polym. Chem.*, **25**, 1653.
8. Pearson, J.M. and Stolka, M. (1981). *Poly(N-vinylcarbazole)*, Gordon and Breach Science Publishers 132.
9. Tao, X., Zhang, Y., Wada, T. and Sasabe, H. (1997). *Appl. Phys. Lett.*, **71**, 1921.
10. Gutsche, C.D. (1991). *Calixarenes: A Versatile Class of Macrocyclic Compounds*, J. Vicens and V. Böhmer (Eds), Kluwer Academic Publishers.
11. Kubo, Y., Tokita, S., Kojima, Y., Osano, Y. and Matsuzaki, T. (1996). *J. Org. Chem.*, **61**, 3758.
12. Clays, K. and Persoons, A. (1993). *Rev: Sci. Instrum.*, **63**, 3285.
13. Marder, S.R., Beratan, D.N. and Cheng, L.-T. (1991). *Science*, **252**, 103.
14. Ishoshima, T., Wada, T., Togo, Y., Kubo, Y. and Sasabe, H. to be published.

8. METAL COMPLEXES WITH FERROMAGNETIC INTERACTIONS

HIROKI OSHIO* and TASUKU ITO

Department of Chemistry, Graduate School of Science, Tohoku University, Aoba-ku, Sendai 980-8578, Japan

ABSTRACT

Strategies of spin polarization mechanism and orthogonal arrangements of the magnetic orbitals are applied to prepare metal complexes with ferromagnetic interactions.

1. INTRODUCTION

Magnetic interactions within or between molecules are typically anti-ferromagnetic. If two magnetic orbitals, each having one electron, are sufficiently close to interact, they can overlap to form antibonding and bonding orbitals. Both electrons locate on the bonding orbital with anti-parallel spin alignment. This situation can be regarded as a very weak chemical bond. However, if the certain conditions are fulfilled, parallel spin alignment of the electrons, that is, ferromagnetic interactions can be obtained. The situations have been well rationalized.[1] (i) Orthogonality of magnetic orbitals. The magnetic interaction J can be expressed as the sum of K (exchange integral) and $2\beta S$ (β: transfer integral and S: overlap integral) which favor the ferromagnetic and antiferromagnetic interactions, respectively. If the magnetic orbitals are (accidentally) orthogonal to each other, the ferromagnetic interaction is obtained.[2] (ii) Configurational mixing of a ground high-spin with charge transfer configurations stabilizes the high-spin ground state.[3] (iii) Topological symmetry of π-electron network applied to design high-spin organic molecules such as poly-carbenes.[4] (iv) Spin polarization, a strategy originally suggested by McConnell,[5] and later documented for inter-molecular stacks having complementary spin alignment, can lead to ferromagnetic interaction.[6] It should be noted that, the existence of direct or indirect charge transfer interactions between the paramagnetic centers can generate or enhance any magnetic interactions, such effects being most conveniently treated with a valence bond-like approach. In this article, the

---

* To whom correspondence should be addressed.

propagation of the ferromagnetic interactions in metal complexes by using following strategies are described. (i) Orthogonal arrangement of magnetic orbitals in connection with charge transfer interactions. (iv) Topological networks of d-π spins of metal ions to organic bridging ligands.

2. ORTHOGONAL ARRANGEMENT OF MAGNETIC ORBITALS

2.1. Metal Complexes with Imino Nitroxide

Magnetic interactions between paramagnetic centers through diamagnetic metal ions are negligibly small or weakly antiferromagnetic. However, some diamagnetic metal ions have been reported to mediate magnetic interactions between coordinated radicals. Diamagnetic metal complexes with semiquinones show a variety of magnetic interactions depending on metal ions and coordination geometries. A series of square planar metal complexes $[M(II)(SQ)_2]$ (M = Ni, Pd, and Pt) (SQ = *tert*-butyl-substituted semiquinone) shows fairly strong antiferromagnetic interactions due to indirect overlap of the magnetic orbitals through the metal d-π orbitals.[7] The strength of inter-radical exchange increases down the series of metal ions; i.e., the strongest coupling for Pt and the weakest for Ni. On the other hand, pseudo-octahedral coordination geometry provides for orthogonal coordination of semiquinones. $[M(III)(3,6\text{-}DBSQ)_3]$ (M = Al and Ga)[8] and $[Ga(III)(3,5\text{-}dtbsq)_3]$[9] showed weak ferromagnetic interactions (J = 6.2, 8.6, and 7.8 cm^{-1} where $H = -2J\Sigma S_i \cdot S_j$), while $[M(IV)(CAT\text{-}N\text{-}SQ)_2]$ (M = Ti, Ge, and Sn) (Cat-N-SQ = tridentate Schiff base biquinone)[10] were characterized by a triplet ground state with the exchange coupling constants of $J = -56$, -27, and -23 cm^{-1} ($H = J\Sigma S_1 \cdot S_2$), respectively. It should be noted that magnitude of magnetic interactions through diamagnetic ions depend strongly on energy level of the d-π orbitals, that is, the charge-transfer interactions plays an important role. In this section, the propagation of magnetic interaction in metal complexes with organic radicals, where the metal ions are diamagnetic, are discussed from the view point of the charge transfer interaction. The organic radicals used are abbreviated as follows.

immepy   impy   bpyim2

2.1.1. Orthogonality with a charge transfer interaction: [Cu(I) (immepy)₂] (PF₆)

Which diamagnetic metal complex provides the appropriate symmetry and orbital energy to propagate ferromagnetic interaction? Ab-initio molecular orbital calculations of the divalent metal oxides have proven that, among the first row transition metal ions, d-orbital energy of the Cu(II) ion is the closest to the oxygen p-orbital,[11] that is, the Cu ion is expected to have the strongest electronic interactions with organic molecules. In addition, Cu(I) ions are known to favor a tetrahedral coordination geometry[12] which is suitable for the orthogonal arrangement of two bidentate ligands. Thus, it is anticipated that coordination of two bidentate radical ligands to the Cu(I) ion would lead to a ferromagnetic interaction between the radicals. The structure of a cation in [Cu(I)(immepy)₂](PF₆) (where immepy is a bidentate imino nitroxide, 2-(2-(6-methylpyridyl))-4,4,5,5-tetramethyl-4,5-dihydro-1*H*-imidazolyl-1-oxy) is shown in Figure 1.[13] Complex molecule has a C_2 axis and the Cu(I) ion is coordinated by the crystallographically equivalent immepy ligands acting as bidentate ligands. The coordination geometry about Cu(I) ion is pseudotetrahedral with the four coordination sites of the Cu(I) ion being occupied by four nitrogen atoms. The Cu-N(imino nitroxide) bond is slightly shorter (1.953(5) Å) than the Cu-N(pyridine) bond (2.081(6) Å). It should be noted that the two radical planes (magnetic orbitals) coordinated to the Cu(I) ion in [Cu(I)(immepy)₂](PF₆) are perpendicular to each other with the dihedral angle of 88.7°.

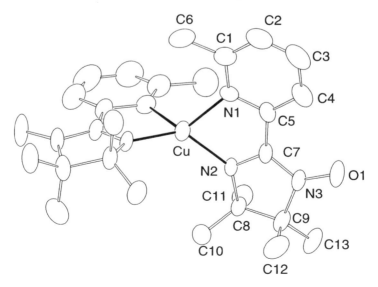

Figure 1   ORTEP view of [Cu(I)(immepy)₂]⁺.

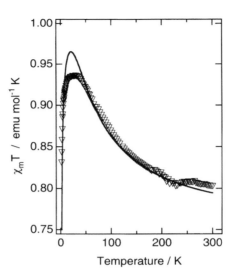

Figure 2 $\chi_m T$ versus T plot of [Cu(I)(immepy)$_2$](PF$_6$).

The magnetic susceptibility data for [Cu(I)(immepy)$_2$](PF$_6$) are shown in Figure 2, in the form of the $\chi_m T$ versus T plot, where χ_m is the molar magnetic susceptibility. The $\chi_m T$ value shows a gradual increase as the temperature decreases, reaching a plateau (0.96 emu mol^{-1}) which is slightly below what would be anticipated for the triplet, and then decreases suddenly below 20 K. This observed temperature dependence of the $\chi_m T$ above 20 K is quite characteristic of a ferromagnetically coupled biradical. The magnetic data permit determination of the triplet-singlet energy gap arising from the intramolecular interaction. The least squares fitting for the data by using a triplet-singlet model ($H = -JS_1 \cdot S_2$)[14] gives the best fit parameters J being 55.1(6) cm^{-1}, where the g value was fixed to be 2.0 and the intermolecular antiferromagnetic interaction ($\theta = -0.6(1)$ K) was included in the calculation.

In acetonitrile solution, [Cu(immepy)$_2$]$^+$ shows intense absorption bands at 766 nm ($\varepsilon = 5000$ M^{-1} cm^{-1}) and 464 nm ($\varepsilon = 6300$ M^{-1} cm^{-1}) with a shoulder band at 510 nm (Figure 3).

The electronic spectra of Cu(I)-diimine, [Cu(I)(TET)]$^+$(TET = 2,2′-bis(6-(2,2′-bipyridyl)biphenyl)[15] and Cu(I) complex with an o-semiquinone ligand, [Cu$^I_2$$\{\mu$-N$_2$[CO(OBu^t)]$_2$[$\mu$-Ph$_2$P(CH$_2$)$_6PPh_2\}_2$]BPh$_4$[16] have been well characterized. The absorption bands for [Cu(immepy)$_2$](PF$_2$) can be assigned by the analogy to the reported data. The intense band at 464 nm and shoulder at 514 nm were assigned to MLCT bands, e(d$_{xz}$, d$_{yz}$) $\rightarrow$ e(π^*) (NLUMO) and b$_2$(d$_{xy}$) $\rightarrow$ e(π^*) transitions, respectively,

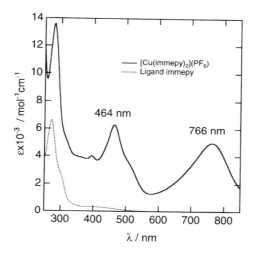

Figure 3 UV-visible spectra of ligand immepy (. . .) and [Cu(I)(immepy)$_2$](PF$_6$) (—) in acetonitrile.

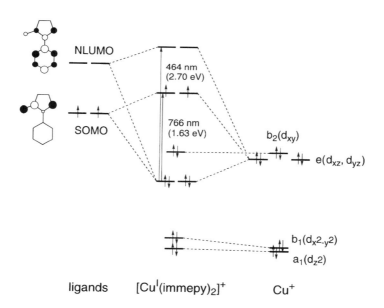

Scheme 1

and the lower energy band (766 nm) would correspond to a e(d$_{xz}$, d$_{yz}$) → SOMO (immepy) transition (Scheme 1).

The mechanism, which determines the relative energy (J) of the triplet and singlet states, has been often discussed in terms of the Heitler-London interaction and valence bond configuration interaction between the ground and charge transfer states.[17] J depends primarily on the Heitler-London type interaction within the ground state and in natural magnetic orbital treatments the triplet-singlet energy gap is expressed as:

$$J_{GS} = 2K_{\pi^*\pi^*} + 4\beta_{\pi^*\pi'}S_{\pi^*\pi^*}$$

where $K_{\pi^*\pi^*}$, $\beta_{\pi^*\pi^*}$ and $S_{\pi^*\pi^*}$ represent the two electron exchange, transfer, and overlap integrals between radical SOMOs, respectively. The two coordinated radicals are orthogonally arranged. As a result, $S_{\pi^*\pi^*}$ is zero and the strong ferromagnetic coupling primarily arises from the two-electron exchange integral, $2K_{\pi^*\pi^*}$. [$Cu^I(immepy)_2$](PF_6) complex, however, showed a fairly strong charge transfer band which corresponds to the π-back donation to the radical SOMO. It is necessary to consider a valence bond configuration interaction to interpret the ferromagnetic interaction. Valence bond-like treatments have been invoked to explain ferromagnetic interactions or ordering in metal complexes as well as in pure organic compounds. For example, Goodenough proposed that interactions between half-filled orbitals on one metal and empty orbitals on the other metal can contribute to ferromagnetic interactions.[18] This mechanism was also invoked to justify the ferromagnetic ordering observed for (p-nitrophenyl)nitronyl nitroxide,[19] and ferromagnetic interactions in [$Mn^{III}(\mu\text{-O})(\mu\text{-}CH_3CO_2)_2Mn^{III}$][20] and [$Gd_2Cu_4$][21] complexes. The situation is not the same as the above case but is closer to the models by McConnel and Breslow.[22] It is assumed that the coordination geometry (D_{2d}) about Cu ion is pseudotetrahedral and use the orthogonal orbitals in the following treatment. Within the four-orbitals (degenerate $e(d_{xz}, d_{yz})$ and degenerate π^* orbitals) and six-electrons (four on $e(d_{xz}, d_{yz})$ and two on each π^*) system, the ground triplet (GS_T) and singlet (GS_S) configurations are represented by

where GS_S has a complement spin configuration. The MLCT configuration generated by the CT transition from e to π^* is represented by

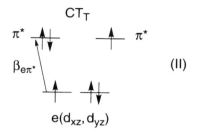

where $\beta_{e\pi^*}$ is a transfer integral. This MLCT state has both triplet (CT$_T$) and singlet (CT$_S$) configurations. However, the triplet is lower in energy than the singlet due to the orthogonality of the singly occupied e and π^* orbitals. The mutual repulsion between the ground and MLCT states leads to a mixing between the two configurations. The admixed triplet and singlet ground states are stabilized by $-2\beta_{e\pi^*}^2/(\Delta - K_{e\pi^*})$ and $-2\beta_{e\pi^*}^2/(\Delta + K_{e\pi^*})$, respectively; where Δ is the cost in energy of transferring an electron from e to π^* and $K_{e\pi^*}$ is the interatomic exchange integral involving the orbitals e and π^*. As a consequence of mixing, the triplet is lower than the singlet and the triplet-singlet energy gap J is

$$J = \frac{4\beta_{e\pi^*}^2 K_{e\pi^*}}{\Delta^2 - K_{e\pi^*}^2}.$$

There is the more excited charge transfer configuration. Starting from the MLCT configuration (II), an electron on a doubly occupied e orbital can be transferred to a singly occupied π^* orbital with the resultant that the Cu ion possesses singly occupied orbitals. By comparison with the GS configuration (I), this corresponds to a double charge transfer configuration and this configuration should be a triplet due to the Hund's rule.

DCT$_T$

π^* ⥮　　⥮ π^*

(III)

↑　　↑

e(d_{xz}, d_{yz})

The interaction between the GS and DCT configurations stabilizes the GS triplet, because the singlet DCT configuration is quite high in energy. The resulting stabilization of the GS$_T$ is $-(\beta_{e\pi^*}^4/\Delta^2)/(E_{DCT} - K_{ee})$, where E_{DCT} and K_{ee} represent the energy cost associated with this electron transfer and the one-site exchange integral, respectively. Stabilization due to the double charge transfer is, in general, very small. In this case, however, the two

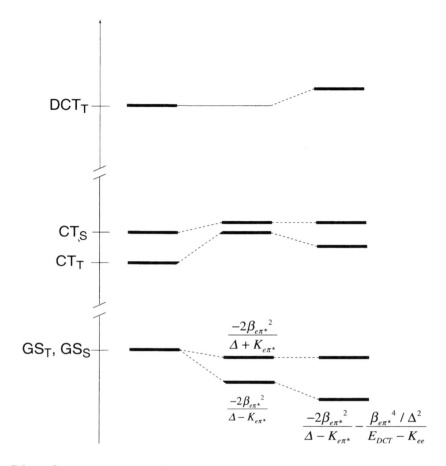

Scheme 2

unpaired electrons locate on the Cu ion in the DCT configuration, which leads to the fairly large exchange integral K_{ee}. Hence, it can be expected that the GS_T-DCT_T configuration interaction can not be ignored. As the result of the configuration interaction of the GS_T with CT_T and DCT_T, the GS_T is stabilized by $-2\beta_{e\pi^*}^2/(\Delta - K_{e\pi^*}) - (\beta_{e\pi^*}^4/\Delta^2)/(E_{DCT} - K_{ee})$.[23] Scheme II summarizes these configuration interactions.

2.1.2. Orthogonality without a charge transfer interaction: [Ag(I)(impy)₂](PF₆)

In [Ag(I)(impy)₂](PF₆) (Figure 4), a coordination geometry about the Ag(I) ion deviates from the tetrahedron with a dihedral angel of the coordinated

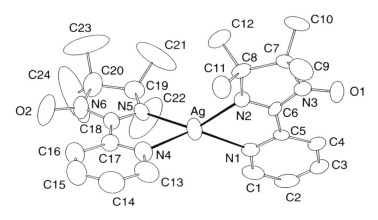

Figure 4 ORTEP view of $[Cu(I)(immepy)_2]^+$.

imino nitroxides of 79.2°, which results in the broken orthogonality of the coordinated imino nitroxides. The average bond lengths between the silver ion and the coordinated nitrogen atoms are 2.375 and 2.228 Å for Ag-N (pyridine) and Ag-N(imino nitroxide), respectively, which are much longer than those of the Cu complex.

The $\chi_m T$ value for $[Ag(I)(impy)_2](PF_6)$ steadily decreased as the temperature was lowered, reaching a minimum value of 0.64 emu K mol^{-1} at 19 K. Below 19 K, a very abrupt rise in $\chi_m T$ is observed and the $\chi_m T$ value reached 0.716 emu K mol^{-1} at 2.5 K. The magnetic interaction between the coordinated imino nitroxide is concluded to be weakly ferromagnetic and this was also confirmed by epr spectra of the frozen methanol solution of the complex, that is, a curie plot of the epr signal intensity for the $\Delta m = 2$ transition gave a positive θ value of 4 K. It should be also noted that $[Ag(I)(impy)_2](PF_6)$ dose not show any metal to ligand charge transfer bands in uv-visible region. Therefore, there is an apparent energy mismatch between the ligand π^* and d-π orbital of the Ag(I) ion. It can be concluded that not only the broken orthogonality, but also the lack of the charge transfer interaction between the Ag(I) ion and iminonitroxides stabilizing the high-spin state, make the intramolecular ferromagnetic interaction very weak in the case of $[Ag(I)(impy)_2](PF_6)$.

2.1.3. Non-orthogonality with a charge transfer interaction: [Pd(II)Cl₂ (immepy)₂]

Reaction of $[NH_4]_2[PdCl_4]$ with immepy gave a square planar Pd(II) complex, $[Pd(II)Cl_2(immepy)_2]$.[24] Two imino nitroxides coordinate to the Pd(II)

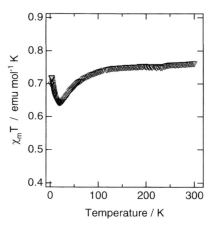

Figure 5 $\chi_m T$ versus T plot of [Ag(I)(impy)$_2$](PF$_6$).

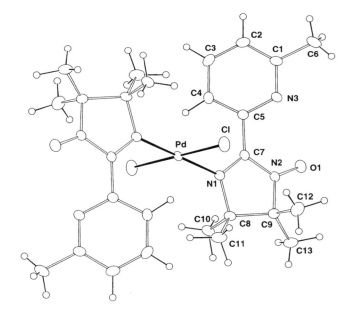

Figure 6 ORTEP view of [Pd(II)Cl$_2$(immepy)$_2$].

ion with the trans-position (Figure 6). The Pd—Cl and Pd—N(imino nitroxide) bond lengths are 2.977(7) and 2.019(2) Å, respectively, and the Cl—Pd—N bond angle is 87.72(5)°.

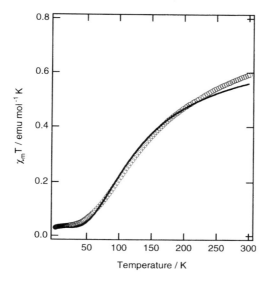

Figure 7 $\chi_m T$ versus T plot of [Pd(II)Cl$_2$(immepy)$_2$].

Magnetic susceptibility measurement shows a fairly strong anti-ferromagnetic interaction being operative between the coordinated imino nitroxides and the exchange coupling constant $2J$ ($H = -2JS_1 \cdot S_2$) was estimated to be -160.8 cm^{-1} with g-value of 2.0 (Figure 7).

A uv-vis spectrum of [Pd(II)Cl$_2$(immepy)$_2$] showed strong absorption bands at 350 nm ($\varepsilon = 7800 \text{ dm}^3 \text{ mol cm}^{-1}$) and 500 nm (1800) (Figure 8).

An electronic spectrum of [PdCl$_2$(bpy)] has been well studied and the band at 317 nm was assigned the MLCT band. A MO calculation of the immepy ligand revealed that the singly occupied molecular orbital (SOMO) is centered mainly on the imino nitroxide moiety, while the next lowest unoccupied orbital (NLUMO) is delocalized over the whole π-system. The band at 300 nm observed in [Pd(II)Cl$_2$(immepy)$_2$] can be, therefore, assigned to be metal to NLUMO and the band at 500 nm corresponds to the metal to SOMO transition. The MLCT interaction implies the metal d$_{xy}$ orbital having a substantial overlap with the SOMOs and lead to the strong anti-ferromagnetic interaction (Scheme 3).

2.1.4. Non-orthogonality without a charge transfer interaction: [Cd(II)(bpyim2) (NO$_3$)$_2$]

In [Cd(II)(bpyim2)(NO$_3$)$_2$] a quadri-dentate biradical ligand, bpyim2, occupies equatorial coordination sites of the Cd(II) ion and two nitrates

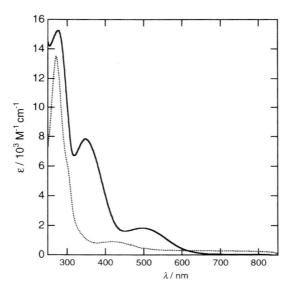

Figure 8 UV-visible spectra of [Pd(II)Cl$_2$(immepy)$_2$] (—) and immepy (…) in methanol.

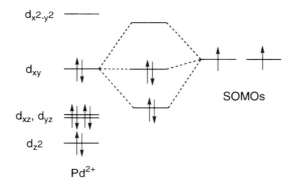

Scheme 3

coordinate to the metal ion from the axial position (Figure 9). The average bond lengths between the cadmium ion and the coordinated nitrogen atoms are 2.407 and 2.429 Å for Cd—N(pyridine) and Cd—N(imino nitroxide), respectively. The complex molecule does not show a MLCT band which might result from the energy mismatch of cadmium d-π orbitals with the ligand SOMO. As expected, the only weak antiferromagnetic interaction

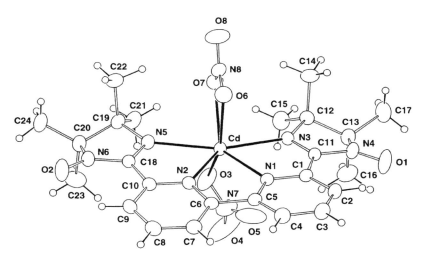

Figure 9   ORTEP view of $[Cd(II)(bpyim2)_2(NO_3)_2]$.

due to the magnetic-dipole interaction are operative between the coordinated radicals.

2.2. $[CrO_4]^{2-}$ bridged metal complexes

The bridging ligands between paramagnetic metal ions usually mediate
antiferromagnetic interactions due to the magnetic orbital overlap through
the bridge. Ferromagnetic interactions, however, can be achieved, if the
magnetic orbitals are (accidentally) orthogonal to each other. In this section,
magnetic interactions through tetraoxo metalates will be discussed by means
of orbital symmetry of the frontier orbitals.

2.2.1. $[CrO_4]^{2-}$ Bridged Cu(II) Complex: $[Cu(II)(acpa)]_2[CrO_4] \cdot 4CH_3$-$OH \cdot 4H_2O$

Some oxometalate anions like $[CrO_4]^{2-}$ have tetrahedral coordination
geometry. In $[CrO_4]^{2-}$ anion, the chromium d-orbitals split into e and t_2 type
orbitals under the T_d symmetry (Scheme 4).

The e pair of the d-orbitals has no matching combination of the oxygen
orbitals, hence, these remain nonbonding. Three t_2 orbitals, however, have
the same symmetry with combinations of coordinated oxygen orbitals, and
form bonding and antibonding orbitals. The t_2-type bonding orbital, mainly
consisting of the oxygen orbitals, are occupied by six electrons from the

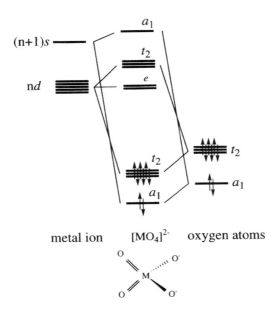

Scheme 4

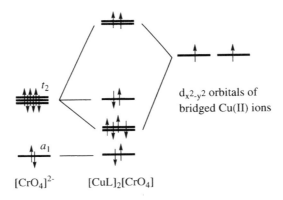

Scheme 5

oxygen atoms. When two paramagnetic metal ions having $d\sigma$ spins like Cu(II) ions are bridged by the $[CrO_4]^{2-}$ unit, two of the t_2-type orbitals and σ-type ($d_{x^2-y^2}$) magnetic orbitals of the Cu(II) ions form bonding and antibonding orbitals, and remainder of the t_2-type orbitals remain non-bonding (Scheme 5). Such dicopper(II) complex bridged by the $[CrO_4]^{2-}$ unit has only mirror symmetry, hence the molecular geometry of the

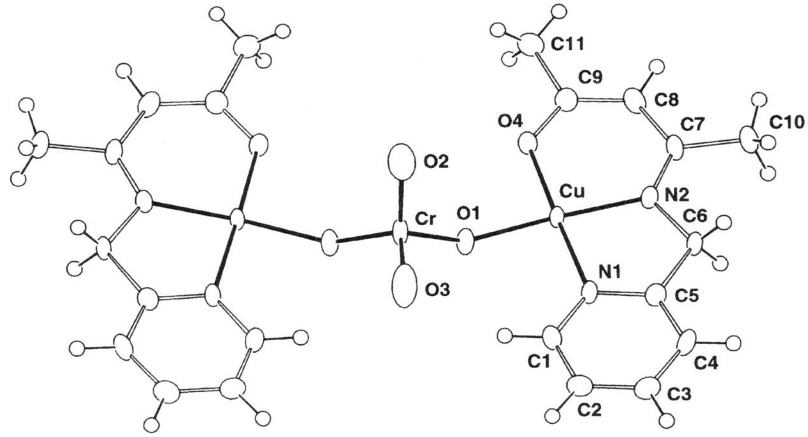

Figure 10 ORTEP view of [Cu(II)(acpa)]$_2$[CrO$_4$].

complex is Cs. In spite of the fact that there are no degenerate orbitals under the Cs symmetry, the formed antibonding orbitals are expected to have a very small energy gap or to be even accidentally degenerate due to the rigid structure of the [CrO$_4$]$^{2-}$ unit. As the results, the two unpaired electrons from the two paramagnetic centers locate on the accidentally degenerate antibonding orbitals, and this leads to the ferromagnetic interaction between the bridged Cu(II) centers.

Reaction of K$_2$[CrO$_4$] with [Cu(acpa)](PF$_6$) (Hacpa = N-1-acetyl-2-propyridene) (2-pyridylmethyl)amine) in water-methanol solution gave a [CrO$_4$]$^{2-}$ bridged dicopper(II) complex, [Cu(acpa)]$_2$[CrO$_4$] · 4CH$_3$OH · 4H$_2$O (Figure 10).

The molecule has the mirror plane on the center of the [CrO$_4$]$^{2-}$ ion which bridges the Cu(II) ions with Cu(II)-Cu(II) separation of 6.443(1) Å. Coordination geometry about the Cu(II) ions is square planar, of which four coordination sites are occupied with N$_2$O chromophore from acpa and oxygen atom from the [CrO$_4$]$^{2-}$ anion and the magnetic orbital of the Cu(II) ion is the d$_{x^2-y^2}$ orbital. The coordination geometry around the Cr(VI) ion is pseudo tetrahedron with the Cr—O bond lengths of 1.606(6)–1.679(4) Å, where the bond lengths with the O1 atoms coordinated to the Cu(II) ion are the shortest among them with bond angles (Cr—O1—Cu) are 141.3(2)° and 149.0(4)°, respectively.

Temperature dependence of magnetic susceptibilities for [Cu(acpa)]$_2$ [CrO$_4$]-4CH$_3$OH · 4H$_2$O is depicted in Figure 11.

The $\chi_m T$ value at 300 K is 0.89 emu mol^{-1} K, which would be expected for the uncorrelated two spins with g value being 2.124. On lowering the

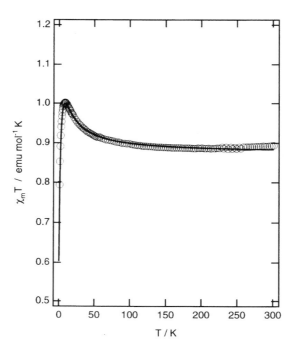

Figure 11 $\chi_m T$ versus T plot of [Cu(II)(acpa)]$_2$[CrO$_4$].

temperature, the $\chi_m T$ value for [Cu(acpa)]$_2$[CrO$_4$] · 4CH$_3$OH · 4H$_2$O increases and exhibits a maximum at 10 K ($\chi_m T = 1.00$ emu mol^{-1} K) and then decreases. The magnetic behavior above 10 K is indicative of the ferromagnetic interaction and data was analyzed by the Bleaney-Bowers equation $(H = -2JS_1 \cdot S_2)$ to give the best fit parameters of $2J = +14.6(1)$ cm^{-1}, $g = 2.12(1)$ and $\theta = -0.8(1)$ K. The ferromagnetic interaction between the Cu(II) ions is predominant at intermediate temperatures and then a weaker antiferromagnetic coupling is involved at lower temperature. The bridging [CrO$_4$]$^{2-}$ ion should be responsible for the ferromagnetic contribution, that is, the tetra-coordination geometry of the [CrO$_4$]$^{2-}$ unit is concluded to arrange the magnetic orbitals of the bridged Cu(II) ions to be orthogonal.[25]

2.2.2. [CrO$_4$]$^{2-}$ Bridged Ni(II) Complex: catena-(μ-CrO$_4$—O,O') [Ni(cyclam)] · 2H$_2$O

The reaction of [Ni(cyclam)](PF$_6$)$_2$ with K$_2$CrO$_4$ in water gave dark red tablets of catena-(μ-CrO$_4$—O,O')[Ni(cyclam)] · 2H$_2$O. Coordination geometry about each Ni(II) ion is octahedral, where the equatorial coordination sites of the Ni(II) ions are occupied by four nitrogen atoms from cyclam (Ni—N = 2.061(2)–2.076(2) Å) and the axial sites are completed by two oxygen atoms from $[CrO_4]^{2-}$ anion (Ni—O = 2.083(2)–2.092(1) Å). $[CrO_4]^{2-}$ anions bridge the Ni(II) ions and form a one-dimensional structure (Figure 12) with the Ni-Cr separation of 3.4100(6)–3.5366(6) Å.

$\chi_m T$ values for *catena*-(μ-CrO$_4$—O,O')[Ni(cyclam)] · 2H$_2$O increase as the temperature is lowered down to 9 K, which is indicative of a ferromagnetic interaction (Figure 13). The Ni—Ni separation within the chain is 6.5954(9) Å, while the closest Ni—Ni separation between the chain is

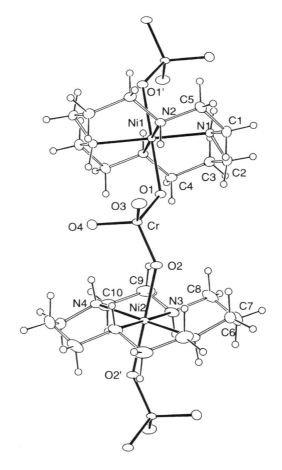

Figure 12   ORTEP view of catena-(μ-CrO$_4$—O,O')[Ni(cyclam)] · 2H$_2$O.

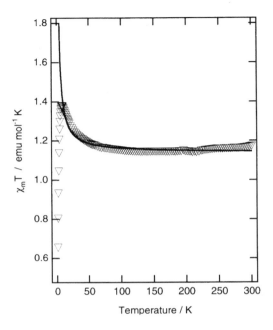

Figure 13 $\chi_m T$ versus T plot of *catena*-(μ-CrO$_4$—O,O′)[Ni(cyclam)] · 2H$_2$O

8.512(1) Å. The observed ferromagnetic interaction is, therefore, due to the intrachain interaction.

The Fisher's model[26] for the classical-spin chain system ($S = 1$ and $H_{\text{chain}} = -J\sum S_i \cdot S_i + 1$) was applied to the analysis of the magnetic data. The magnetic susceptibility χ_m can be expressed as

$$\chi_m = \frac{Ng^2\beta^2 S(S+1)}{3kT}\frac{1+u}{1-u}$$

$$u = \coth\left[\frac{JS(S+1)}{kT}\right] - \left[\frac{kT}{JS(S+1)}\right],$$

where the symbols have their usual meaning. The least-squares fitting of the observed data above 9 K led to $J = +0.6(1)$ cm^{-1} and $g = 2.13(1)$. Sudden decrease of $\chi_m T$ values below 9 K might be due to an antiferromagnetic interchain interaction.

Optical conductivity spectra of *catena*-(μ-CrO$_4$—O,O′)[Ni(cyclam)] · 2H$_2$O, which were obtained by the Kramers-Kronig transformation from the polarized reflectivity spectra for the single crystal, were depicted in Figure 14. When the electric vector E is parallel to the chain (E // chain), a

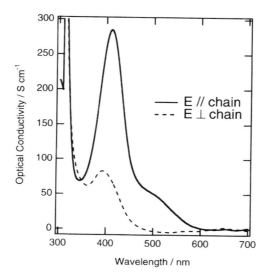

Figure 14 Optical Conductivity spectra of *catena*-(μ-CrO$_4$—O,O′)[Ni(cyclam)] · 2H$_2$O for the single crystal with $E /\!/$ chain (—) and $E \perp$ chain (- - -).

strong absorption band (410 nm) with a shoulder (510 nm) was observed, where the former band at 410 nm was assigned to the LMCT(O → Cr) band[27] and the later shoulder band was assigned to the LMCT band from the coordinated oxygen atom to the Ni(II) ion.[28] Due to the LMCT(O → Ni) interaction, the spin on the d_z2 orbital is considered to delocalize onto the coordinated oxygen atom of [CrO$_4$]$^{2-}$ anion.

Propagation of the ferromagnetic interaction in *catena*-(μ-CrO$_4$—O,O′) [Ni(cyclam)] · 2H$_2$O can be understood by the orbital topology of the frontier orbitals. Under the T_d symmetry [CrO$_4$]$^{2-}$ ion has triply degenerate HOMOs (highest occupied molecular orbital) being combinations of t_2-type chromium d and coordinated oxygen p orbitals. When [CrO$_4$]$^{2-}$ anion bridges Ni(II) ions from their axial sites, each having two spins on d_z2 and $d_x2_{-y}2$ orbitals, the d_z2 of the Ni(II) ions and two of the triply degenerate orbitals of [CrO$_4$]$^{2-}$ fragment form two sets of σ-type bonding and anti-bonding orbitals. The d_z2 orbitals of the Ni(II) ions and the oxygen p orbitals interact to mix through the LMCT(O → Ni) interaction. The two unpaired electrons from the d_z2 orbitals occupy the antibonding orbitals, while the $d_x2_{-y}2$ and remainder of [CrO$_4$]$^{2-}$ orbitals remain nonbonding (Scheme 6). The two antibonding MOs having two spins must be energetically close enough to stabilize the high-spin state (ferromagnetic interaction).

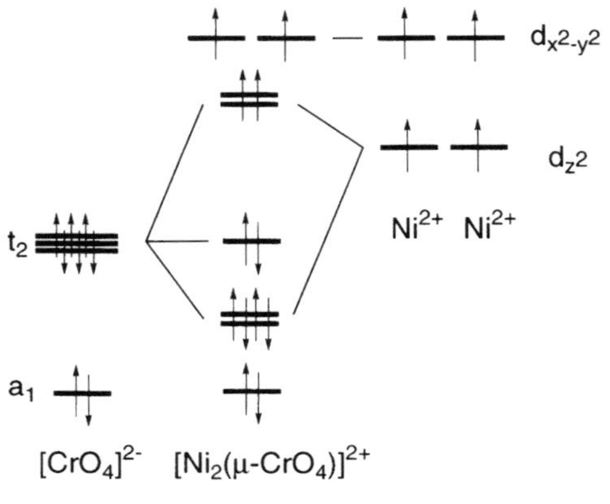

Scheme 6

3. TOPOLOGICAL APPROACH

Some strategies for incorporation of ferromagnetic interactions in organic multi-radical compounds have been proposed.[29] A spin polarization mechanism, i.e., topological symmetry of the p electron network, was applied to design high-spin organic molecules in poly-carbene system. Some poly-carbenes, in which the carbenes are linked in meta-positions of a bezene-ring, have shown fairly strong ferromagnetic interaction and the ferromagnetic interaction between radicals can be explained by polarized p-π spin over the whole molecule which forms a topological network (see below).

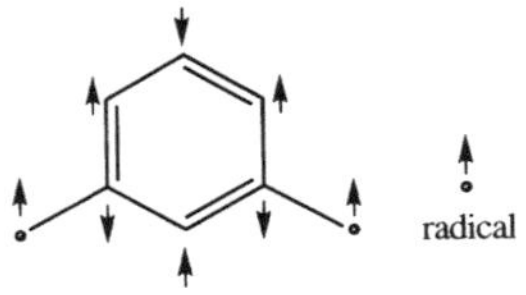

Is it possible to apply the spin polarization mechanism to designing multi-nuclear complexes with an intramolecular ferromagnetic interaction? That is, if the paramagnetic metal ions in multinuclear systems are linked with an aromatic bridging ligands in the meta-position each other, the intramolecular ferromagnetic interaction can be propagated due to the topologically networked d-π spins to the ligand p-π orbitals. In order to have the ferromagnetic interaction, some conditions should be fulfilled. (i) Central metal ions should have d-π spins. (ii) Organic bridging ligands should have de-stabilized HOMOs which are close in energy to the magnetic orbitals of the metal ions. (iii) A topological network concerning the d-π spin to the ligand p-π orbitals must exist. A bridging ligand designed, which satisfies these condition, is 4,6-bis[N,N-bis(2'-pyridylmethyl)aminomethyl]-2-methyl-resorcinol abbreviated as H_2bpmar. A dinuclear Fe(III) complex, [Fe$_2$(bpmar)(H$_2$O)$_4$](NO$_3$)$_4 \cdot$ 3H$_2$O, in which central metal irons are high-spin d^5, was prepared and the magnetic properties were characterized.

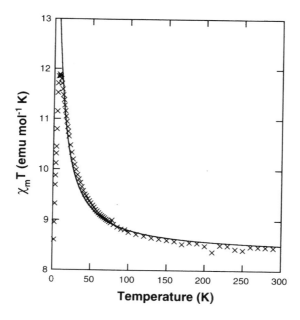

bpmar^{2-}

H$_2$bpmar

The $\chi_m T$ vs. T plot (Figure 15) for [Fe$_2$(bpmar)(H$_2$O)$_4$](NO$_3$)$_4 \cdot$ 3H$_2$O shows a gradual increase as the temperature decreased with a maximum $\chi_m T$ value (11.88 emu mol^{-1} K) being reached at 7.0 K. This magnetic

Figure 15 $\chi_m T$ versus T plot of [Fe$_2$(bpmar)(H$_2$O)$_4$](NO$_3$)$_4 \cdot$ 3H$_2$O.

behavior indicates a ferromagnetic interaction between iron centers. An abrupt decrease of the $\chi_m T$ value below 7.0 K might be due to an inter-molecular antiferromagnetic interaction.

The ferromagnetic exchange coupling constant was estimated to be $J = +0.65(3)\,cm^{-1}$ and $g = 1.953(4)$ by using the data above 10 K $(H = -2JS_1 \cdot S_2)$.

It is apparent from the view point of the symmetry that the metal ion should have d-π spin in order to interact or mix with the organic p-π orbital. Simple perturbation theory predicts that two orbitals can strongly interact only if the orbitals have the same symmetry and comparable in energies. It is assumed that the the d-π orbitals is higher in energy than the p-π orbitals of the organic molecule. There are two ways to mach the orbital energy of the two frontier orbitals. First, a metal ion, of which energy level of the d-π orbital is comparable with that of p-π orbital of the organic bridges, should be selected. Second, bridging ligands, whose frontier orbitals have a higher energy, should be available for this purpose. The d-orbitals of the first row transition metal ions are the lowest in energy among the transition metal ions and the iron ion has d-π orbitals with rather lower energy than the other first row transition metal ions. Thus, the iron ion is expected to have the d-π orbital which is closer in energy to the p-π orbitals of the organic molecule. On the other hand, the bridging ligands, of which HOMO has relatively high energy, can be prepared by chemical modifications of organic bridging ligands. The HOMO of an organic molecule can be stabilized by the introduction of electron attracting groups like halogen atoms, while the introduction of electron donating groups or negative charges to a molecule results in the de-stabilization of the HOMO. Chemical modifications of the bridging ligands seems to be easier way to adjust the orbital energies to the metal d-π orbitals.

Resorcinol, the bridging group of $[Fe_2(bpmar)(H_2O)_4](NO_3)_4 \cdot 3H_2O$ having two negative charges, has a de-stabilized HOMO (2.00 eV) due to the charges on the molecule. It is expected that the d-π orbitals of the iron ions easily mix with p-π orbitals of the resorcinol. The magnetic susceptibility experiment of the iron(III) complex reveals a ferromagnetic interaction between iron centers ($J = +0.65\,cm^{-1}$). The spin polarization of the d-π spin to the bridging ligands operates to have the ferromagnetic interaction. If we assume an α-spin on the iron ion, a β-spin on the carbon atom is induced through the p-π electrons of the coordinating oxygen atom. This β spin is polarized as an α spin on the adjacent carbon atoms, and so on. Finally an a-spin on the other side of the iron atom is induced, that is, the ferromag-netic interaction between the two iron centers is propagated (Scheme 7).

Cu(II) dinuclear complex, $[Cu_2(bpmar)(NCS)_2] \cdot 4H_2O$, in which has the same bridging ligand as $[Fe_2(bpmar)(H_2O)_4](NO_3)_4 \cdot 3H_2O$ and Cu(II) ions do not have any d-π spin, does not show a ferromagnetic interaction.

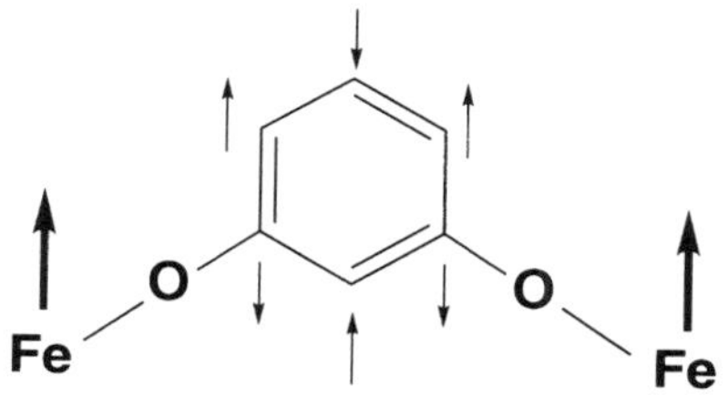

Scheme 7

It should be noted from this experiment that the d-π spin in $[Fe_2(bpmar)(H_2O)_4](NO_3)_4 \cdot 3H_2O$ plays an important role leading to ferromagnetic interactions.[30]

4. CONCLUSION

Control of intermolecular interactions is a key for constructing magnetically ordered system, Orthogonal arrangement of magnetic orbitals, which include and relate to charge transfer interactions is the easiest and efficient way to have a ferromagnetic interactions. Extension of the orthogonality to the 3D system promises the construction of molecular based ferromagnet.

REFERENCES

1. a) Kollmar, C. and Kahn, O. (1993). *Acc. Chem. Res.*, **26**, 259. b) Miller, J.S. and Epstein, A.J. (1994). *Angew. Chem. Int. Ed. Engl.*, **33**, 385.
2. (a) Kahn, O., Galy, J., Journaux, Y., Jaud, J. and Morgenstern-Badarau, I. (1982). *J. Am. Chem. Soc.*, **104**, 2165. (b) Pei, Y., Journaux, Y. and Kahn, O. (1989). *Inorg. Chem.*, **28**, 100. (c) Caneschi, A., Gatteschi, D., Laugier, J. and Rey, P. (1987). *J. Am, Chem, Soc.*, **109**, 2191. (d) Tamaki, H., Zhong, Z.J., Matsumoto, N., Kida, S., Koikawa, M., Achiwa, N., Hashimoto, Y. and Okawa, S. (1992). *J. Am. Chem. Soc.*, **114**, 6974. (d) Oshi, H. and Nagashima, U. (1992). *Inorg. Chem.*, **31**, 3295.
3. (a) Miller, J.S., Calabrese, J.C., Rommelmann, H. Chittipeddi, R., Zhang, J.H., Reiff, W.M. and Epstein, (1987). *J. Am. Chem. Soc.* (109), 769. (b) Turek, P., Nozawa, K., Shimoi, D., Awaga, K., Inabe, T., Maruyama, Y. and Kinoshita, M. (1991). *Chem. Phys. Lett.*, **180**, 327.
4. (a) Itoh, K. (1967). *Chem. Phys. Lett.*, **1**, 235. (b) Sugawara, T., Bandow, S., Kimura, K. Iwwamura, H. and Itoh, K. (1986). *J. Am. Chem. Soc.*, **108**, 368. (c) Teki, Y., Takuti, T., Itoh, K. Iwamura, H. and Kobayashi, K. (1986). *J. Am. Chem. Soc.*, **108**, 2147. (d) Fujita, I., Teki, Y., Takui, T. Knoshita and T. Itoh, K. (1990). *J. Am. Chem. Soc.*, **112**, 4074.
5. McConnel, H.M. (1963). *J. Am. Chem. Soc.*, **39**, 1910.
6. (a) Miller, J.S., Epstein, A.J. and Reiff, W.M. (1988). *Acc. Chem. Res..*, **21**, 114. References therein. (b) Kollmar, C., Vouty, M. and Kah (1991). *J. Am. Chem. Soc.*, **113**, 7994.

(c) Izuoka, A., Murata, S., Sugawara, T. and Iwamura, H. (1987). *J. Am. Chem. Soc.*, **109**, 2631.

7. (a) Fox, G.A. and Pierpont, C.G. (1992). *Inorg. Chem.*, **31**, 3718. (b) Abakumov, G.A., Cherkasov, V.K., Bunov, M.P., Ellert, O.G., Raitin, U.V., Zakharov, Struchkov, Y.T. and Saf'yanov, U.N. (1992). *Isv. Akad. Nauk SSSR.*, 2315.

8. Lange, C.W., Conklin, B.J. and Pierpont, C.G. (1994). *Inorg. Chem.*, **33**, 1276.

9. (a) Adams, D.M., Rheingold, A.L., Dei, A. and Hendrickson, D.N. (1993). *Angew. Chem. Int. Ed. Engl.*, **32**, 391. (b) Ozarowski, A., McGarvey, B.R., El-Hadad, A., Tian, Z., Tuck, D.G., Krovich, D.J. and DeFtis, G.C., (1993). *Inorg. Chem.*, **32**, 841.

10. Bruni, S., Caneschi, A., Cariati, F., Delfs, C. Dei, A. and Gatteschi, D. (1994). *J. Am. Chem. Soc.*, **116**, 1388.

11. Yamaguchi, K. Nakano, M, Namimoto, H. and Fueno, T. (1998). *Jpn. Appl. Phys.*, **27**, L1835.

12. Cotton and Wilkinson, G. (1988). *Advanced Inorganic Chemistry 5th edition*, Wiley, New York.

13. Oshio, H., Watanabe, T., Ohto, A., Ito, T. and Nagashjima (1994). *Angew. Chem. Int. Ed. Engl.*, **33**, 670.

14. Bleaney, N. and Bowers, K.D. (1952). *pro Soc. London, Ser. A*, **214**, 451.

15. Müller, E., Piguet, C., Bernardinelli, G. and Williams, A.F. (1988), *Inorg. Chem.*, **27**, 849.

16. a) Moscherosch, M., Field, J.S., Kaim, W., Kohlmann, S., Krejcik, M. (1993). *J. Am. Chem. Soc. Dalton Trans*, 211. b) Vogler, C., Hausen, H.-D., Kaim, W., Kohlmann, S., Kramer, A. and Rieker, J. (1989). *Angew. Chem.*, **101**, 1734.

17. a) Kahn, O. *Molecular Magnetism*; VCH Publishers: Weinheim, New York, 1993. b) Seggern, I., Tuczek, F. and Bensch, W. (1995). *Inorg. Chem.*, **34**, 5530. c) Tuczek, F. and Solomon (1994). *J. Am. Chem. Soc.*, **116**, 6916.

18. Goodenough, J.B. (1963). *Magnetism and Chemical Bond*; Interscience: New York.

19. Tamura, M., Nakazawa, Y., Shiomi, D., Nozawa, K., Hosokoshi, Y., Ishioka, M., Takahashi, M. and Kinoshita, M. (1991). *Chem. Phys. Lett.*, **86**, 401.

20. Hotzelmann, R., Wieghardt, K., Flörke, U., Haupt, H.-J., Weatherburn, D.C., Bonvoisin, J., Blondin, G. and Gierard, J-J. (1992). *J. Am. Chem. Soc.*, **114**, 1681.

21. Andruh, M., Ramade, I., Codjovi, E., Guillou, O., Kahn, O. and Trombe, J.C. (1993). *J. Am. Chem. Soc.*, **115**, 1822.

22. (a) McConnel, H.M. (1976). *Proc. Robert A. Welch Found. Cof. Chem. Res.*, **11**, 144. (b) Breslow, R., Juan, B., Kluttz, R.Q. and Xia, C.Z. (1982). *Tetrahedron*, **38**, 863. (c) Breslow, R. (1982). *Pure Appl. Chem.*, **54**, 927.

23. (a) Tuczek, F. and Solomon, E.I. (1993). *Inorg. Chem.*, **32**, 2850. (b) Shen, Z., Allen, J.W., Yeh, J.J., Kang, J.-S., Ellis, W., Spicer, W., Lindau, I., Maple, M.B., Dalichaouch, Y.D., Torikachvili, M.S., SuZ. and Geballe, T.H. (1987). *Phys. Rev. B.*, **36**, 8414.

24. Oshio, H., Ohto, A. and Ito, T. (1996). *Chem. Commun.*, 1541.

25. Oshio, H., Kikuchi, T. and Ito, T. (1996). *Inorg. Chem.*, **35**, 4938.

26. Fisher, M.E. (1964). *A Phys.*, **32**, 343.

27. a) Duiker and Ballhausen, C.J. (1968). *Theoret. Chim. Acta.*, **12**, 325. b) Johnson, L.W. and McGlynn, S.P. (1970). *Chem. Phys. Lett.*, **7**, 618. c) Miller, R.M., Tinti, D.S. and Case, D.A. (1989). *Inorg. Chem.*, **28**, 2738.

28. Oshio, H., Okamoto, H., Kikuchi, T. and Ito, T. (1997). *Inorg, Chem.*, **36**, 3201.

29. (a) McConnel, H.M. (1963). *J. Chem. Phys.*, **31**, 299. (b) Mataga, N. (1968). *Theor. Chim. Acta*, **10**, 372. (c) Ovchinnikov, A.A. (1978). *Theor. Chim. Acta*, **47**, 297. (d) Itoh, K. (1978). *Pure Appl. Chem.*, **50**, 1251. (e) Breslow, R., Juan, B., Kluttz, R.Q. and Xia, C.Z. (1982). *tetrahedron*, **38**, 863.

30. (a) Oshio, H. and Ichida, H. (1995). *J. Phys. Chem.*, **99**, 3294. (b) Oshio, H. (1991). *J. Chem. Soc. Chem. Comm.*, 1227.

9. PREPARATION AND ASYMMETRIC ORIENTATION OF DIPOLAR DENDRONS IN AN ASSEMBLED STRUCTURE

SHIYOSHI YOKOYAMA* and SHINRO MASHIKO

Communications Research Laboratory, 588-2 Iwaoka, Nishi-ku, Kobe 651-2492, Japan

INTRODUCTION

This paper describes: i) the synthesis and characterization of rod-shaped dendrons based on an electron donor/acceptor azobenzene chromophore; ii) the preparation of their assembled structures by using the Langmuir-Blodgett (LB) film transfer technique, and iii) the characterization of the molecular structure, especially the molecular orientation and polar ordering of the assembled thin films, by second harmonic generation (SHG) measurement. The goal of this study is to advance novel type rod-shaped dendrons as the basis of a dipolar dendritic molecular system.

Artificial preparation techniques for supermolcular structures are important to learn how to fabricate highly organized materials that can be used for future electronic and optoelectronic applications. In organic and polymeric materials, these is a great ability to self-organize with the formation of molecular aggregations or supermolecular structures with properties different from those of disordered structures (Philp, 1991; Lehn, 1988; Lindsey, 1991). Therefore, there have been many studies to control molecular structures and orientations. A technique to prepare self-assembled monolayers and Langmuir-Blodgett (LB) films has been shown to have great potential in the fabrication of such highly organized organic structures (Roberts, 1990; Ulmann, 1991). In these films, chromophoric units are highly ordered by means of alkyl chains and arranged in respect to the surface normal. An important advantage of using these techniques is that one can precisely fabricate two or three-dimensional molecular assemblies and extend these to more complex architectures (Katz, 1991; Penner, 1994).

In terms of the recent advances in supermolecular architecture in organic and polymeric materials, dendritic maclomolecules, called "*dendrons or dendrimers*", have attracted a lot of interest (Ardoin, 1995; Issberner, 1994, Fréchet, 1994; Tomalia, 1990; Tomalia, 1994). They have unique physical

---

* To whom correspondence should be addressed.

and chemical properties, which are very different from one-dimensional linear polymers in terms of size, shape, and conformation. One of the most peculiar characteristics of dendritic macromolecules is their compactness (Mourey, 1992), which means they have a practical application in achieving the high constituent density in functional molecular devices. Although earlier dendrimer research has targeted globular structures, rod-shaped dendrons or dendrimers, which can be arranged to form the assembled structure, are of potential interest in the field of electronic and optoelectronic applications. Our target compounds are rod-shaped dendrons. When such dendrons have chromophoric branching with an electron donor/ acceptor system, the materials become dipolar and their electronic asymmetry is interesting especially from the viewpoint of nonlinear optical activity. In order to obtain rod-shaped and dipolar dendrons, we selected an azobenzene chromophore as the repeat branching. The azobenzene derivative used in this study has a rod-shaped structure, and it has a π-electron conjugation coupled with electron donor and acceptor groups. The technique we used to prepare the molecularly organized thin films was the LB film transfer method.

We characterized the structure of products by ^1H and ^{13}C NMR spectroscopy, size exclusion chromatography (SEC), and the matrix assisted laser desorption ionization time of flight mass spectroscopy (MALDI TOFMS). The thin films formed on the glass substrates were characterized by the second harmonic generation (SHG) technique, which is very sensitive to the molecular asymmetry of hyperpolarized molecules arranged without a center of inversion (Shen, 1989). The measurement provided us with not only structural information but also the potential ability of the products as the nonlinear optical material.

PREPARATION OF DENDRONS

Scheme 1–4 show the synthesis strategy of the dipolar dendrons (Yokoyama, 1997), which is referred to as a *convergent* approach (Hawker, 1990; Hawker, 1992). Through a judicious choice of reagents and synthesis routes, one can precisely introduce the functional molecules and the building blocks at a defined location. The starting compound is the 4-carboxy-4'-[bis(2-hydroxyethyl)amino]azobenzene **1**, which was obtained after the diazo-coupling reaction of 4-aminobenzoic acid and N-phenyldiethanolamine under acidic condition. This compound was modified with an electron donor and acceptor groups, and with two OH groups and one acid group as the AB$_2$ type monomer. Compound **2**, which was in our case the first generation, was obtained by the reaction of **1** with hexadecanoyl chloride in the presence of pyridine as the base. The alkyl chains in

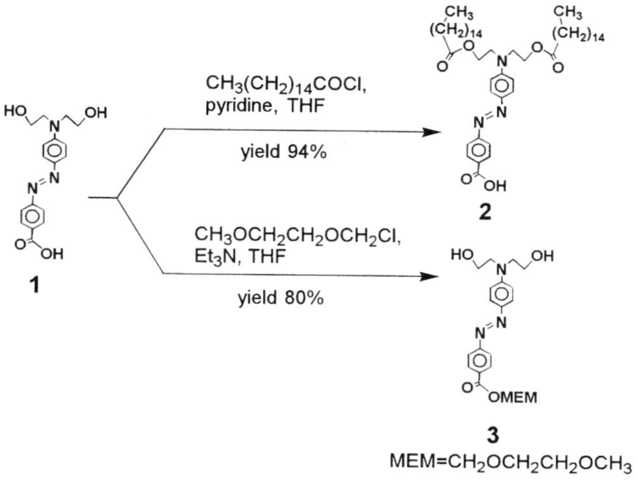

Scheme 1

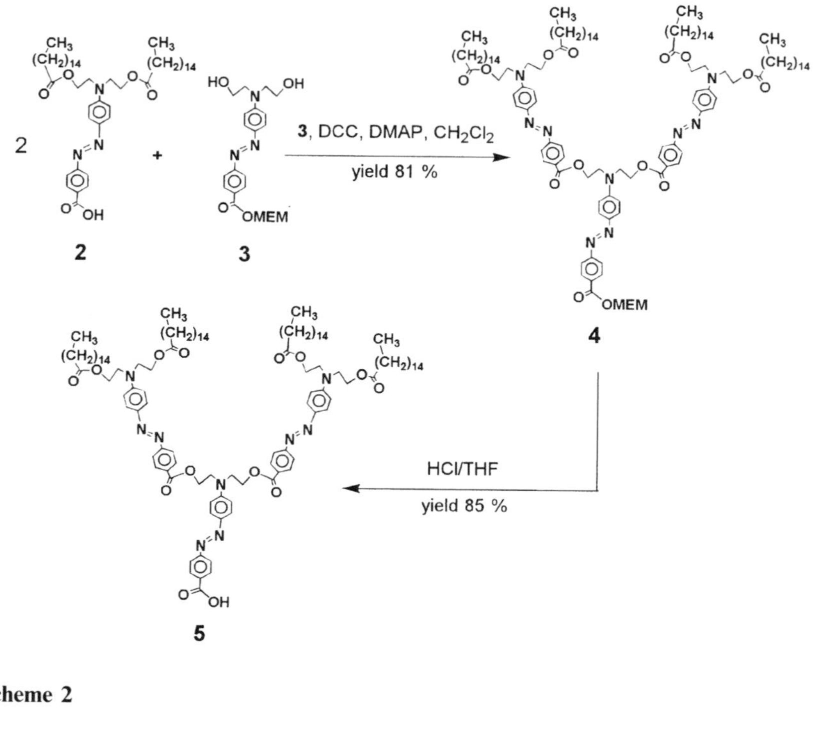

Scheme 2

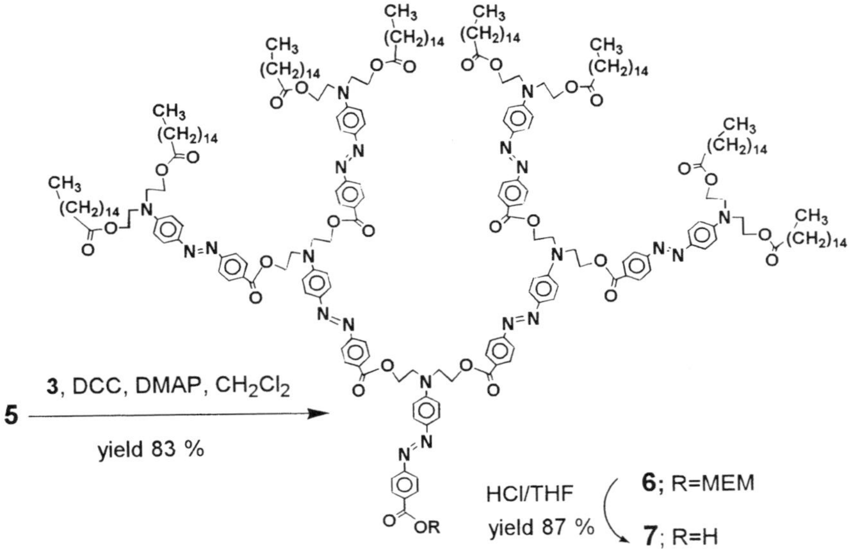

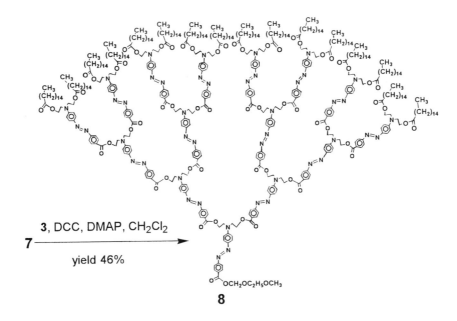

5 $\xrightarrow{\text{3, DCC, DMAP, CH}_2\text{Cl}_2}$ yield 83 %

HCl/THF
yield 87 %

6; R=MEM

7; R=H

Scheme 3

7 $\xrightarrow{\text{3, DCC, DMAP, CH}_2\text{Cl}_2}$ yield 46%

8

OCH$_2$OC$_2$H$_5$OCH$_3$

Scheme 4

compound **2** will be an exterior unit in the final dendrons. A methoxy-ethoxymethyl (MEM) unit was used to protect the carboxyl group of **3**, where the reaction of **1** and MEM chloride in the presence of triethylamine produced compound **3**. The key step in dendron synthesis was the coupling reaction of monomers, which was optimized by using dicyclohexyl-carbodiimide (DCC) and dimethylaminopyridine (DMAP) in dichloromethane (Klausener, 1972; Höfle, 1978). The reaction of **2** and **3** proceeded quantitatively at room temperature to produce compound **4**, generation 2. Another important reaction was the removal of MEM protection. Since MEM ester is sensitive to acidic conditions, the deprotection reaction of compound **4** was carried out in 2–4% HCl/THF solution at below 5 °C. The product is compound **5**, having acid at the focal point. Repeating the above two steps of, coupling and deprotection reactions using the same reagents generates the growth of dendrons. The reaction of **5** and **3** produced dendron **6**, generation 3, and subsequently the corresponding acid **7**. Similarly dendrons **8**, generation 4, was obtained after coupling **7** and **3**. All the dendrons were easily soluble in various solvents such as dichloromethane, chloroform, ethyl acetate, and THF, and could be precipitated in polar solvents such as methanol. The compounds synthesized were purified by column chromatography or HPLC, where the large R_f difference in silica gel between compounds with MEM ester at the focal point and those with an acid group allowed easy purification.

The structure of the products was thoroughly characterized by ^{1}H and ^{13}C NMR spectroscopy. Figure 1 and 2 show the selected NMR spectra for dendrons **6** and **7**. For dendron **6**, the MEM group at the focal point provided characteristic resonances in the ^{1}H NMR spectrum at 3.4, 3.6, 3.9, and 5.6 ppm, and in the ^{13}C NMR spectrum at 59.1, 69.7, 90.0, and 71.5 ppm. An integration of the resonance due to MEM protons in the ^{1}H NMR spectrum and a comparison with other resonances confirmed the exact degree of branching, or generation number, as expected. Removal of the MEM group was confirmed by the disappearance of these resonances in ^{1}H and ^{13}C NMR spectra, and by a shift in the focal point carbonyl resonance from 165.6 to 169.5 ppm in the ^{13}C NMR spectrum. Similar NMR results were obtained in dendrons **4**, **5**, and **8** (not shown).

The most important characterization in dendron synthesis is the high purity and the monodispersed molecular weight of the products. SEC has become a widely used technique to characterize polymeric systems, and has been extended to the study of dendritic macromolecules in particular. Figure 3 shows a series of SEC traces for compounds **4–8**. Each compound has a narrow peak, from which we determined they were 99.5% pure. The experimental molecular weights, Mn and Mw, of these compounds were determined by comparing their SEC traces with those of the standard materials of monodisperse polystyrenes. The results are summarized in

(a)

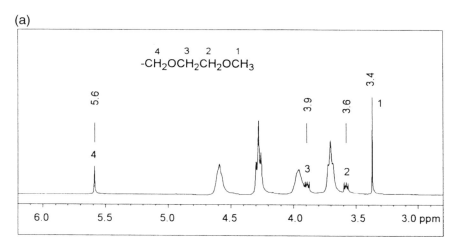

(b)

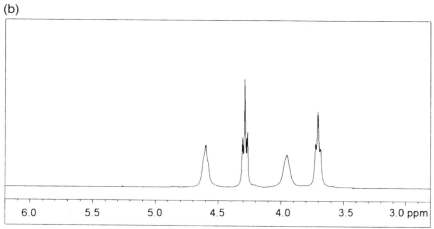

Figure 1 Selected ¹H NMR spectra (in CDCl₃) of (a) compound **6** and (b) compound **7**. Insert is protonic assignments of MEM unit.

Table 1. Although SEC had high purity levels and narrow monodispersities, Mw/Mn, of molecular weights, dendrons had significant errors in Mn and Mw as expected for their structures. Since SEC is a measure of the hydrodynamic volume of solutes and not of the molecular weight, the large differences in Mn and Mw compared with those of standard polystylenes suggest the unique structure and conformation of dendrons modified by flexible alkyl chains at the exterior and relatively rigid azobenzene units in branching.

(a)

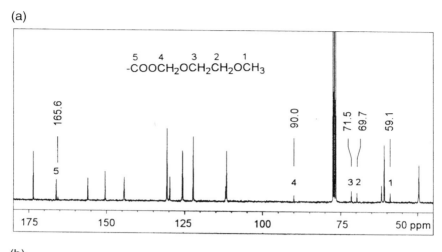

(b)

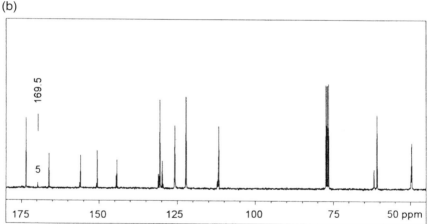

Figure 2 Selected ^{13}C NMR spectra (in CDCl$_3$) of (a) compound **6** and (b) compound **7**. Insert is carbon assignments of MEM unit.

The exact molecular weights of the dendrons were confirmed by MALDI TOFMS. Figure 4 shows a typical MALDI TOFMS spectrum for compound **8**. The spectrum has two intense peaks at m/z = 8598 and 8620, which are assigned to [M + H]$^+$ and [M + Na]$^+$ ionized peaks, respectively. The additional small peaks observed in the smaller mass region are not impurities, but are due to the fragmented peaks produced during measurement. The m/z values of dendrons **4–8** are summarized in Table 1. All the experimental results were identical with the correct molecular weights as

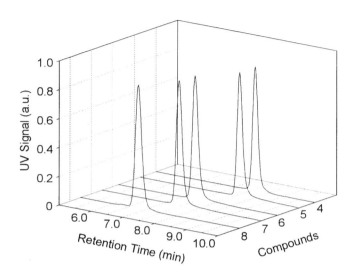

Figure 3   SEC traces of compounds **4**, **5**, **6**, **7**, and **8** in THF after purification.

Table 1   Molecular weight characterizations of dendrons by SEC and MALDI TOFMS.

Compound	Molecular Weight (calculated)	SEC			MALDI TOFMS
		Mw	Mn	Mw/Mn	m/z (calcd. for $[M + H]^+$)
4	1995.0	3523	3340	1.05	1997
5	1906.9	3435	3528	1.03	1907
6	4194.1	7312	6887	1.06	4199
7	4106.0	6949	6752	1.03	4112
8	8592.4	12762	12040	1.06	8598

expected for the structure. The removal of MEM protection at the focal point in dendrons was evident from in MALDI TOFMS spectra, where the decrease in m/z of *ca.* 90 from compound **4** to **5**, and **6** to **7**, corresponded to the mass of a MEM unit.

PREPARATION OF ASSEMBLED STRUCTURES

The synthesized dendrons had an amphiphilic structure, where alkyl chains were located at the exterior position and azobenzene repeats were altered in the branches. Therefore, such a structure paved the way to construct

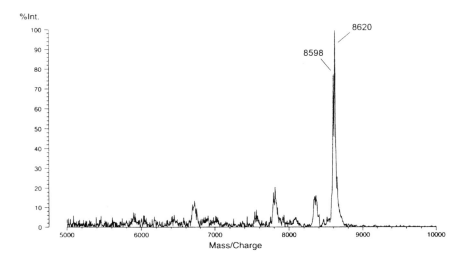

Figure 4 MALDI TOFMS spectrum of compound **8**.

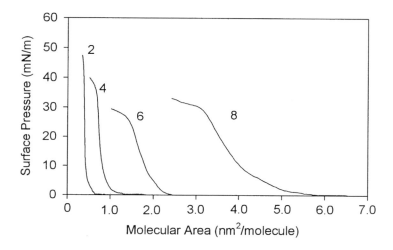

Figure 5 Surface pressure and molecular area curves of compounds **2**, **4**, **6**, and **8** on pure water surface in a Langmuir trough.

assembled films using the Langmuir-Blodgett (LB) film transfer technique. All the compounds were dissolved in chloroform with a concentration of 1–0.05 mmol/l, and spread on a pure water surface in a Langmuir trough. Figure 5 shows the surface pressure and area curves of compounds **2**, **4**, **6**,

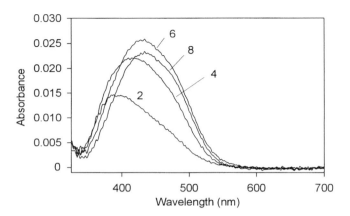

Figure 6   Absorption spectra of thin films of **2**, **4**, **6**, and **8**. Films were prepared on the both side of glass substrates.

and **8** on the water surface. Compound **2** has a steep increase in the surface pressure, where the extrapolated molecular area (to zero surface pressure) was $0.42 \, nm^2$/molecule. Since this value was identical with the sum of the cross-section of two alkyl chains, the surface film was expected to be closely packed analogous to conventional LB materials. In such a film, alkyl chains are highly ordered and a chromophore unit should be arranged with respect to the surface normal. The molecular areas increased as the generation of dendrons increased. This change supported the expectation that the coupling reactions would double the number of exterior alkyl chains and the size of dendrons. Compounds **6** and **8** had a gradual increase in the surface pressures and they formed rather expanded films on the water surface.

All the films could be transferred from the water surface onto a glass substrate at a constant surface pressure of $20 \, mN/m$. Figure 6 shows the absorption spectra for the films, **2**, **4**, **6**, and **8**. All the films had a characteristic band of π-π^* electron transition of azobenzene unit at 380–430 nm. In the film of **2** the band shifted to the shorter wavelength. This is a typical tendency of LB film, in which molecules are closely arranged to form a J-type structure. As the generation of dendrons increases, the band shifted to a longer wavelength. The band at 430 nm in the film of **8** is identical with the band of the azobenzene chromophore in solution.

SHG OF DENDRONS IN ASSEMBLED THIN FILMS

Recently a great deal of research has described the characterization techniques for molecular orientation or polar ordering in organic thin films. In

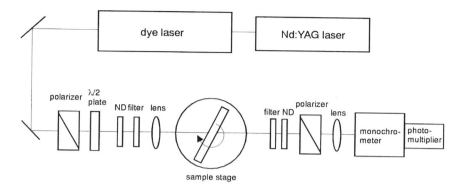

Figure 7 Schematic illustration of the optical set-up used to examine SHG in the thin films.

particular, the surface measuring of second harmonic generation (SHG) is a sensitive proving technique in the study of surfaces and interfaces, and it has been used extensively to study organic adsorbates such as Langmuir films, Langmuir-Blodgett monolayer or multilayer films, and self-assembled monolayer films (Marowsky,1988; Ashwell, 1992; Lupo, 1988; Ashwell, 1994; Heinz, 1982). Thus, the SHG technique is especially useful in the present study to characterize the assembled structure and dipolar properties of dendrons in thin films. The experimental equipment for the SHG measurements is shown schematically in Figure 7. The output of a dye laser at 1064 nm pumped by a Q-switched Nd-YAG laser was used as the fundamental beam. Pulse energies of < 1.0 mJ were used for all measurements on the sample. The sample was set on a rotation stage to vary the incident angles of the fundamental beam. A y-cut quartz single crystal ($d_{11} = 0.32$ pm/V) was used as a reference sample. In SHG measurement, Langmuir-Blodgett monolayer films have been widely used for the model structure, where chromophoric units are oriented by means of alkyl chains without a center of inversion. We used this model and assumption in order to estimate the second-order nonlinear coefficients and molecular orientation of the films. Details on theories and calculations are described in other references (Ashwell, 1992; Zhang, 1990; Katz, 1991; Marowsky, 1990).

For SHG measurements, pure dendron films or diluted films in the matrix of arachidic acid LB film were prepared on glass substrates. We compared SHG properties in the pure films and diluted films. The diluted films were prepared by mixing dendrons with arachidic acid in chloroform solution, and then spreading them on a water surface. The diluted films were very stable on the water surface compared with pure films, and they had higher surface collapse pressures to form the stable surface films.

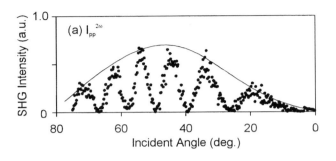

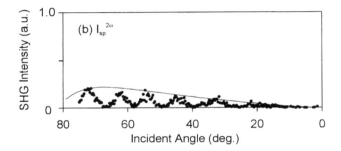

Figure 8 SHG intensity as a function of fundamental beam incident angle from a glass substrate covered with the film of **2**. (a) p-Polarized SHG fringe by p-polarized fundamental beam ($I_{pp}^{2\omega}$) and (b) s-polarized SHG fringe by p-polarized fundamental beam ($I_{sp}^{2\omega}$). The solid lines are theoretical envelope fitted to the data.

Figure 8 shows the SHG fringes obtained in the film of **2**. The fringes were obtained by measuring the p-polarized SHG intensity by p and s-polarized fundamentals as a function of the incident angles relative to the sample plane. The fringes resulted from the phase difference between two SHG waves generated at either side of the films on the glass. The complementary constructive and destructive interference indicates that the SHG is fully coherent, which is characteristic of a film with uniform polar ordering or molecular orientation. In the film of **2**, we set up a simple model to determine the second order nonlinear optical coefficients. In this film we assumed a model structure analogous to conventional SHG active LB monolayer films, where the molecule had one main component of second order molecular hyperpolarizability along the molecular axis and had an average tilt angle relative to the surface normal with a random azimuthal distribution. Since the optical properties of azobenzene chromophores are primarily governed by π electrons in a system with two phenyl groups passing through nitrogen atoms, the molecules can be treated as being rod-shaped and their

Table 2 SHG results for dendron films.

Compound	Normalized $I_{pp}^{2\omega}/I_{quartz}(\times 10^{-4})^a$	
	in pure film	in assembled film[b]
2	0.01	0.2
4	0.6	5.3
6	6.0	11.8
8	1.7	63.7

[a] SHG intensity ($I_{pp}^{2\omega}$) is relative to a quartz reference
[b] Dendrons are diluted in the matrix film of arachidic acid

transition dipole is parallel to the molecular axis. The theoretical curves in Figure 8 are envelopes fit to the SHG data. The best fit between the theoretical and the experimental envelope gave second-order nonlinear coefficients of $d_{zzz} = 16.4$ pm/V and $d_{zxx} = 8.9$ pm/V. The average tilt angle of the chromophore was calculated to be 47° relative to the surface normal. It should be noted that similar characters of SHG fringes were obtained in films for dendrons **2**, **4**, and **6**, though they have multiple chromopholic branches. Their SHG intensities increased as the generation of the dendrons increased (see Table 2). This suggests that these dendrons behave like rod-shaped molecules and are arranged at the glass interface without a center of inversion. However, a significant decrease in the SHG was noted in the film of **8**, suggesting that the dendron could not form an ordered structure presumably due to their spreading branching and expanded film morphology.

In order to determine the effect of a dendritic structure on the formation of assembled structures and their SHG activities, dendrons were diluted in the matrix of arachididc acid LB monolayer films. Using dendron **8**, diluted films were systematically prepared and their SHG properties were determined. Figure 9 shows the surface pressure and area curves for dendron **8** mixed with arachidic acid of various ratios. The diluted films were much more stable on a water surface compared with pure films. The surface pressure and area curves showed a steep increase in the surface pressues up to 60–70 mN/m. As dendrons were diluted with excessive arachidic acid, a phase transition in surface films could be observed around 30 mN/m. The films were transferred from the water surface onto glass substrates at 20 mN/m.

Figure 10 shows the change in SHG intensities relative to a quartz reference versus fractional dye densities in the diluted films. The fractional dye density is the relative surface concentration of the azobenzene unit in the diluted film compared to that in the pure film. The SHG intensities initially

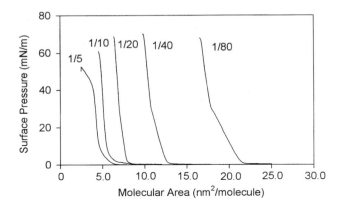

Figure 9 Surface pressure and molecular area curves of compound **8** with arachidic acid on pure water surface in a Langmuir trough. The mixing ratios of **8**/arachidic acid are 1/5, 1/10, 1/20, 1/40, and 1/80.

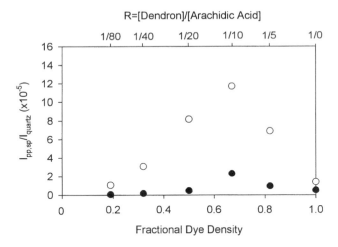

Figure 10 Plot of the SHG intensities I_{pp} (○) and I_{sp} (●) relative to the SHG intensity of a quartz reference versus the fractional dye density of azobenzene unit in the diluted films.

increase with decreasing fractional dye density. The plot shows a maximum at $R = 1/10$ (where $R = $ [dendron]/[arachidic acid], in mol), where SHG activity is most efficient due to optimum molecular orientation and high chromophore density. In diluted chromophoric LB films SHG is generally

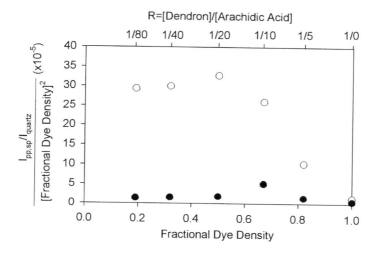

Figure 11 Plot of the SHG intensities I_{pp} (○) and I_{sp} (●) normalized to a coverage concentration of 100%, relative to the SHG intensity of a quartz reference versus the fractional dye density of azobenzene unit in the diluted films.

scaled with the square of molecular density (Giring, 1987). Thus, experimental SHG intensities were normalized to a coverage concentration of 100% as shown in Figure 11. The scaled results are characterized by a plateau region extending between $R = 1/10-1/80$. In these films the dendrons had a constant film structure and molecular orientation. A similar effect of dilution on SHG activity was observed in the films of **2**, **4**, and **6**.

Table 2 summarizes the SHG intensities of dendrons in pure films and diluted films relative to the SHG intensity of a quartz reference. For a relative comparison of the SHG activity, experimental SHG intensities were normalized by dividing the square of fractional densities of dendrons in the films. For all dendrons, SHG activity in the diluted films was found to be superior to that in pure films. This suggests that the good organizational ability of arachidic acid significantly improves molecular orientation and packing arrangement by means of alkyl chains. The significant enhancement of SHG in the film of **8** is noteworthy, where the SHG activity was most efficient. Since normalized SHG activity is directly linked to the polar parameter of dendrons, which includes molecular hyperpolarizability, local field factors, and molecular orientation, the SHG results indicate that the polar ordering of dendrons increases as generation increases. The second-order nonlinear optical coefficients in these films were estimated to be $d_{zzz} = 15-27 \, \text{pm/V}$ and $d_{zxx} = 8-15 \, \text{pm/V}$. These observations support our synthesis strategy that the introduction of both alkyl chains and rod-shaped branching units into dendrons functionalized products, so that they became rod-shaped material and novel non-linear optical material.

CONCLUSIONS

We made progress toward our goal of preparing novel dipolar dendritic materials based on electron donor/acceptor chromophoric branching. This progress was a result of: i) the stepwise synthesis of dendritic structures modified with azobenzene branching and alkyl chain exteriors; ii) the preparation of assembled structures by using the LB film transfer technique; and iii) the characterization of asymmetrically arranged dendrons by using a sensitive surface proving technique for SHG measurement. We demonstrated the versatile application of present dendritic structures as second order nonlinar optical material. The experimental results indicated that the introduction of a rod-shaped azobenzene derivative as the branching is an ingenious way of making the dendritic structure uniaxial, and there is particular merit in this structure in generating SHG. The behavior of alkyl chains at the exterior of dendrons is also unique in the formation of assembled structures. Our achievement in the present study is a combination of controlled material synthesis, the fabrication technique for molecular assemblies, and a sensitivity proving technique. We believe that such a technological combination would be especially useful in fabricating more complex supermolecular systems for future applications.

REFERENCES

Ardoin N. and Astruc, D. (1995). Molecular Trees: From Synthesis Towards Applications, *Bull. Soc. Chim. Fr.*, **132**, 875.

Ashwell, G.J., Hargreaves, R.C., Baldwin, C.E., Bahra, G.S. and Brown, C.R. (1992). Improved Second-Harmonic Genaration From Langmuir-Blodgett Films of Hemicyanine Dyes, *Nature*, **357**, 393.

Ashwell, G.J., Jackson, P.D. and Crossland, W.A. (1994). Non-Centrosymmetry and Second Harmonic Generation in Z-Type Langmuir-Blodgett Films, *Nature*, **368**, 438.

Fréchet, M. (1994). Functional Polymers and Dendrimers: Reactivity, Molecular Architecture and Interfacial Energy, *Science*, **263**, 1710.

Giring, I.R., Cade, N.A., Kolinsky, P.V., Jones, R.J., Peterson, I.R., Ahmad, M.M., Neal, D.B., Petty, M.C., Roberts, G.G. and Feast, W.J. (1987). Second-Harmonic Generation in Mix Hemicyanine: Fatty-Acid Langmuir-Blodgett Monolayers, *J. Opt. Soc. Am. B.*, **4**, 950.

Hawker, J. and Fréchet, J.M. (1990). Preparation of Polymers with Controlled Molecular Architecture. New Convergent Approach to Dendritic Macromolecules, *J. Am. Chem. Soc.*, **112**, 7638.

Hawker J. and Fréchet, J.M. (1992). Unusual Macromolecular Architectures: The Convergent Growth Approach to Dendrite Polyesters and Novel Block Copolymers, *J. Am. Chem. Soc.*, **114**, 8405.

Heinz, F., Chen, C.K., Ricard, D. and Shen, Y.R. (1982). Spectroscopy of Molecular Monolayers by Resonant Second-Harmonic Generation, *Phys. Rev. Lett.*, **48**, 478.

Höfle, G., Steglich, W. and Vorbrüggen, H. (1978). 4-Dialkylaminopyridines as Highly Active
Acylation Catalysts, *Angew. Chem. Int. Ed. Engl.*, **17**, 569.

Issberner, J., Moors, R. and Fögtle, F. (1994). Dendrimers: From Generation and Functional
Groups to Functions, *Angew. Chem. Int. Ed. Engl.*, **33**, 2413.

Katz, H.E., Scheller, G., Putvinski, T.M., Schilling, M.L., Wilson, W.L. and Chidsey, C.E.D.
(1991). Polar Orientation of Dyes in Robust Multilayers by Zirconium Phosphate-Phos-
phonate Interlayers, *Science*, **254**, 1485.

Klausener Y.S. and Bodansky, M. (1972). Coupling Reagents in Peptide Synthesis, *Synthesis*,
453.

Lehn, J.M. (1988). Supermolecular Chemistry-Scope and Perspectives Molecules, Supermolecules and Molecular devices, *Angew. Chem. Int. Ed. Engl.*, **27**, 89.

Lindsey, S. (1991). Self-Assembly in Synthetic Routes to Molecular Devices. Biological Principles and Chemical Perspectives, *New. J. Chem.*, **15**, 153.

Lupo, D., Prass, W., Scheunemann, U., Laschewsky, A., Ringsdorf, H. and Ledoux, I. (1988).
Second-Harmonic Generation in Langmuir-Blodgett Monolayers of Stilbazium Salt and
Phenylhydrazone Dyes., *J. Opt. Soc. Am. B.*, **5**, 300 (1988)

Marowsky G. and Steinhoff, R. (1988). Hemicianine Monolayer Orientation Studied by Second
Harmonic Generation, *Opt. Lett.*, 13, 707.

Marowsky, G., Lüpke, G., Steinhoff, R., Chi, L.F. and Möbius, D. (1990). Anisotropic Second-
Order Nonlinearities of Organic Monolayers, *Phys. Rev. B.*, **41**, 4480.

Mourey, H., Turner, S.R., Rubinstein, M., Fréchet, J.M.J., Hawker, C.J. and Wooley, K.
(1992). Unique Behavior of Dendritic Macromolecules: Intrinsic Viscosity of Polyether
Dendrimers, *Macromolecules*, **25**, 2401.

Penner, T.L., Motschmann, H.R., Armstrong, N.J., Ezenyilimba, M.C. and Williams, D.J.
(1994). Efficient Phase-Matched Second-Harmonic Generation of Blue Light in an Organic
Waveguide, *Nature*, **367**, 49.

Philp D. and Stoddart, J.F. (1991). Self-Assembly in Organic System, Synlett, 445.

Roberts, G. (1990). *Langmuir-Blodgett Films*, Plenum Press: New York.

Shen, R. (1989). Surface Properties Probed by Second-Harmonic and Sum-Frequency Generation, *Nature*, **337**, 519.

Tomalia, D.A., Naylor, A.M. and Goddard III, W.A. (1990). Starburst Dendrimers: Mole-
cular-Level Control of Size, Shape, Surface Chemistry, Topology and Flexibility from Atoms
to Macroscopic Matter, *Angew. Chem. Int. Ed. Engl.*, **29**, 138.

Tomalia, A. (1994). Starburst/Cascade Dendrimers: Fundammental Building Blocks for a New
Nanoscopic Chemistry Set, *Adv. Mater*, **6**, 529.

Ulmann, A. (1991). An Introduction to Ultrathin Organic Films From Langmuir-Blodgett to
Self-Assembly, Academic Press: San Diego, CA.

Yokoyama, S., Nakahama, T. and Mashiko, S. (1997). Functional Dendritic Macromolecules:
Preparation and Optical Properties, *Mol. Cryst. Liq. Cryst.*, **294**, 19.

Yokoyama, S., Nakahama, T., Otomo, A. and Mashiko, S. (1997). Preparation and Assembled
Structure of Dipolar Dendrons Based Electron Donor/Acceptor Azobenzene Branching,
Chem. Lett., 1137.

Zhang, C.H. and Wong, G.K. (1990). Determination of Molecular Orientation in Molecular
Monolayers by Second-Harmonic Generation, *J. Opt. Soc. Am. B.*, **7**, 902.

10. ANISOTROPY OF OPTICAL RESPONSES IN TWO-DIMENSIONAL MOLECULAR SYSTEMS

TAKASHI ISOSHIMA*, TATSUO WADA
and HIROYUKI SASABE

*Frontier Research Program, The Institute of Physical and Chemical Research
(RIKEN), and Core Research for Evolutional Science and Technology
(CREST), Japan Science and Technology Corporation (JST)
2-1 Hirosawa, Wako, Saitama 351-0198, Japan*

ABSTRACT

Anisotropy in the linear and nonlinear optical properties of a molecular system is discussed in terms of the "dimensionality" of the component molecule. Two categories of optical response in two-dimensional (2-D) molecules are investigated, and it is shown that they present an anisotropy different from ordinary one-dimensional (1-D) molecules. The first category is a resonant nonlinear optical process in a molecule with 2-D degenerate transitions. Anisotropy of transient photoinduced absorption change in a metallophthalocyanine with C_{4v} symmetry is investigated as a model. It is derived that the polarization dependence in pump-probe response is 4:3 (less anisotropic than a 1-D molecule) or 8:1 (more anisotropic), depending on the nature of the transition for probe whether it carries the information of pump polarization or not. This polarization dependence of 4:3 is experimentally demonstrated by subpicosecond pump-probe spectroscopy. The second category is a nonresonant linear optical process in a molecule with nondegenerate 2-D transitions. Anisotropy of refractive index in a poled polymer based on a disubstituted carbazole with 2-D charge transfer (CT) character is investigated as a model. In this molecule, it is theoretically predicted that the molecular polarizability is much more isotropic than in its 1-D CT counterpart, resulting in smaller anisotropy and smaller change in refractive index due to orientation relaxation. Advantage of this feature in photonic device applications is discussed in terms of tolerance to orientation relaxation. Experimental demonstration of this feature of small anisotropy is presented.

---

* Tel.: +81-48-462-1111 ext. 3456, Fax: +81-48-467-9389, E-mail: isoshima@postman.riken.go.jp

1. INTRODUCTION

In order to design a molecule with a desired function, it is essential to understand the structure-to-property relation. One of the most distinct example is the relation between anisotropic properties and dimensionality of the molecular structure: anisotropy in linear and nonlinear optical properties is governed by the symmetry of the molecule through the dimensionality of its electronic structure. In hyper-structured molecular design, the dimensionality should be considered in the component groups and in the whole molecule. A hyper-structured molecule such as a dendrimer is composed of various functional groups coupled to each other. Therefore, the dimensionality of each functional groups contributes the optical anisotropy of the whole molecule. At the same time, it is known that a dendrimer presents a two-dimensional (planar) structure at a lower generation and evolves to a three-dimensional (globular) structure at a higher generation.[1] Hence, the relation between optical anisotropy and dimensionality must be understood in order to design a hyper-structured molecule with a desired photonic function.

Anisotropy in nonlinear optical response is also very important in terms of material characterization and application. For example, polarization dependence of an all-optical switch is governed by nonlinear optical anisotropy, and in order to realize a polarization-independent all-optical switch less anisotropy of nonlinear optical response is preferable. Optical Kerr-gates[2] utilize nonlinear optical anisotropy, and thus it is necessary to know the polarization dependence for evaluation of the third-order nonlinear optical coefficient components from the experimental result. Anisotropic refractive index in an oriented molecular system such as poled polymer is very important in photonic device applications. For example, in some organic photorefractive materials, anisotropy of refractive index (i.e., birefringence) due to molecular orientation is dominant in formation of a refractive index grating.[3] In a directional coupler for electro-optic modulation, the coupling length is affected by the refractive index. In an optical second harmonic generation (SHG) device, the phase matching condition is extremely sensitive to the refractive index. Orientation relaxation in a poled polymer affects refractive index through anisotropy of molecular polarizability, resulting in a fatal degradation of device performance. It is therefore of great significance to investigate a molecular system in terms of dimensionality in order to control and optimize the optical anisotropy.

In this article we discuss the relation between two-dimensionality and optical anisotropy. Two categories of two-dimensionality are described: the anisotropy in resonant optical response due to two-dimensionally degenerate transitions, and the anisotropy in nonresonant optical response due to nondegenerate two-dimensional (2-D) transitions. Here the term "two-dimensional" is used when an optical process is dominated by two transition dipole moments in different direction. As the first category of two-dimensionality, we discuss the anisotropy in photoinduced transient absorption change of a metallophthalocyanine, which has four-fold symmetry and thus two-dimensionally degenerate transitions. As the second category, we discuss the anisotropy in refractive index induced by poling of a disubstituted carbazole, which has 2-D intramolecular charge-transfer (CT) transitions.

2. TRANSIENT ABSORPTION CHANGE IN A METALLOPHTHALOCYANINE: A RESONANT PROCESS IN A DEGENERATE 2-D SYSTEM

Randomly oriented molecules with one-dimensional (1-D) character are known to present a polarization dependence of 3:1 in the photoinduced change of absorption coefficient and refractive index, if the depolarization due to rotational molecular reorientation is not considered. On the other hand, molecules with 2-D symmetry such as metallophthalocyanines and with three-dimensional symmetry such as C_{60} present a different anisotropy in their nonlinear optical response. Polarization dependence of fluorescence in a 2-D system has been investigated by several groups,[4–7] and it was shown that the polarization dependence is initially 8:1 (i.e., fluorescence anisotropy $r = 0.7$) and relaxes to 4:3 (i.e., $r = 0.1$).[6,7] Anisotropy in a pump-probe experiment was also investigated to some extent,[7,8] and the same polarization dependence was concluded. Recently, it has been shown that the initial polarization ratio in transient absorption change is governed by the nature of probe transition, and 4:3 ratio can be observed as an initial polarization dependence.[9]

A metallophthalocyanine can be a model compound for this investigation, since it presents optical transitions of twofold spatial degeneracy (hereafter called "2-D transitions") due to its fourfold symmetry of C_{4v} or D_{4h},[10–12] as shown in Figure 1. Phthalocyanines are well studied materials in various fundamental and application research fields[10,13] including optical limiting and switching[14,15] and enhanced optical nonlinearities in excited states,[16,17] because of their stability and large optical nonlinearity. Additionally, metallophthalocyanines can accept a diversity of substitutions for central metals and peripherals, which enables us to tune a variety of properties such as eigenenergy and solubility. Therefore they can also be introduced to a hyper-structured molecule as functional core and/or peripheral components.

Here we analyze the anisotropy in the photoinduced transient absorption change originating from 2-D transitions. We demonstrate experimentally

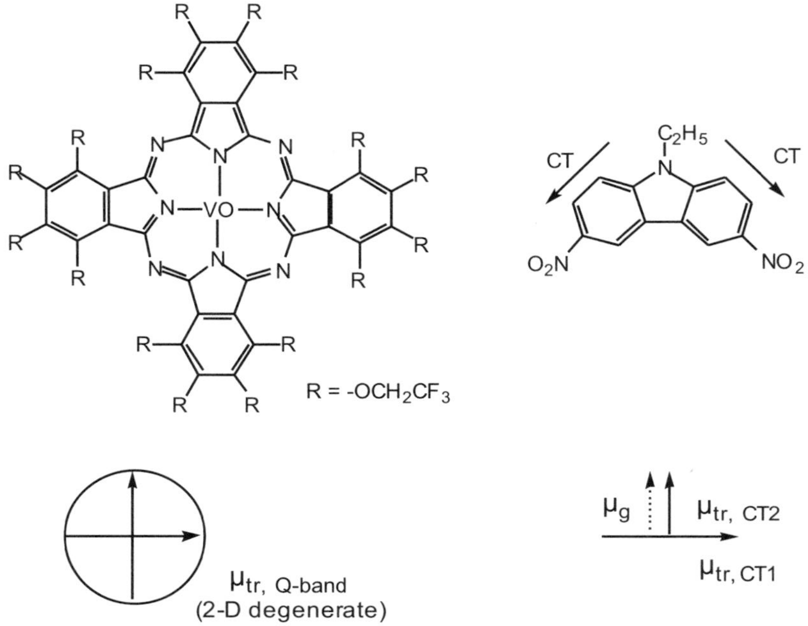

Figure 1 Molecular structure of a 2-D degenerate molecule VOPc(TFE)$_{16}$ (left) and a 2-D CT molecule 3,6-dinitro-N-ethylcarbazole (right). Arrangement of 2-D transition dipole moments is presented schematically at the bottom.

this anisotropy with a solution of vanadylphthalocyanine derivative with C_{4v} symmetry by means of subpicosecond pump-probe spectroscopy.[9,18]

2.1. Theory

Anisotropy in the photoinduced transient absorption change is governed by two projection factors: one is between the polarization of the excitation (pump) light and the transition dipole which responds to the pump; the other is between the polarization of the probe light and the transition dipole which responds to the probe. In a randomly oriented molecular system such as a liquid or solid solution, the total response is obtained by averaging the projection factors over the solid angle. Assume that the molecule is planar and that all transitions are π-π^* transitions which are 2-D and in the molecular plane. π-π^* transitions in a 2-D molecule with C_{4v} or higher symmetry are two-fold degenerate and polarized in the molecular plane, since π-π^* transitions are allowed only between a twofold degenerate

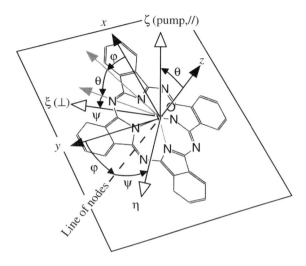

Figure 2 Definition of Eulerian angles (φ, ψ, θ) between the molecular coordinate system (x, y, z) and the laboratory coordinate system (ξ, η, ζ).

e_g-symmetric orbital and a nondegenerate orbital with a_1, a_2, b_1, or b_2 symmetry. We also assume that the highest occupied molecular orbital (HOMO) is nondegenerate and the lowest unoccupied molecular orbitals (LUMOs) are degenerate. In order to describe the projection factors, hereafter the Eulerian angles (φ, θ, ψ) are defined for the transformation from the molecular coordinate system (x, y, z) to the laboratory coordinate system (ξ, η, ζ), as shown in Figure 2. In the molecular coordinate system, the x and y axes are defined to be in the molecular plane and the z axis normal to it. In the laboratory coordinate system, the η direction is defined as the propagation direction of the pump and probe beam and the pump polarization in the ζ direction. Thus the probe polarization is either in the ζ or ξ direction. Hereafter the former case is called "parallel" configuration since the pump and probe polarizations are parallel to each other, and the latter case "perpendicular" configuration since they are perpendicular to each other.

Due to the degeneracy of states in the plane of the molecule, the linearly polarized pump will result in a superposition of the degenerate excited states so that the projection factor is maximized. Defining the x-axis in the molecular coordinate system as the excitation axis, the first Eulerian angle φ is always 0 and the projection factor in the plane of the molecule is always 1. Therefore, the projection factor due to the pump, p_{pump}, is:

$$p_{pump} = \sin^2 \theta. \tag{1}$$

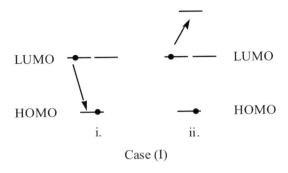

Case (I)

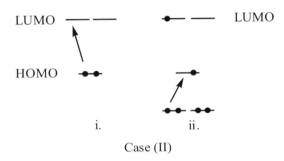

Case (II)

Figure 3 Schematic energy diagram of molecular orbitals and typical transitions of the model molecule. Case (I): examples of transitions which reflect the anisotropy in the molecular plane due to excitation. i: stimulated emission. ii: excited-state absorption *from* the degenerate orbital. Case (II): examples of transition whose probability is isotropic in the molecular plane. i: bleaching (which is the same transition as the ordinary ground-state absorption). ii: excited-state absorption *to* the nondegenerate orbital from a lower-energy one.

On the other hand, the probe projection factor depends on three Eulerian angles. It also depends on the optical transition concerned. Assuming a molecule with a nondegenerate HOMO and two-dimensionally degenerate LUMOs, there are typically two cases as shown in Figure 3: (I) the electron already excited by the pump is promoted to another orbital and thus the transition reflects the anisotropy in the plane due to the excitation; (II) an electron different from the one excited by the pump is promoted to another orbital and thus the transition probability is isotropic in the molecular plane. Examples of case (I) are the stimulated emission from the excited state to the ground state (case (I)-i in Figure 3) and the excited-state absorption in which the electron is excited from the degenerate orbital to another upper one (case (I)-ii). Note that in this probe transition the final orbital must be nondegenerate. In this case, the promoted electron carries the information of the pump polarization, and thus the probe projection factor for the parallel polarization configuration, $p_{\text{probe I}//}$, is

$$p_{\text{probe I}//} = \sin^2 \theta \tag{2}$$

and that for the perpendicular polarization configuration, $p_{\text{probe I}\perp}$, is

$$p_{\text{probe I}\perp} = \cos^2 \theta \sin^2 \psi \tag{3}$$

Hence, the average of the total projection factor for the random orientation is

$$\langle p_{\text{pump}} \, p_{\text{probe I}//} \rangle = \langle \sin^4 \theta \rangle = \frac{8}{15} \quad \text{Parallel configuration} \tag{4}$$

$$\langle p_{\text{pump}} \, p_{\text{probe I}\perp} \rangle = \langle \sin^2 \theta \cos^2 \theta \sin^2 \psi \rangle = \frac{1}{15} \quad \text{Perpendicular configuration} \tag{5}$$

where $<>$ indicates average over the solid angle. Therefore the parallel-to-perpendicular ratio is 8:1. Spontaneous emission from the excited state to the ground state can be treated in the same way as depicted in Case (I)-i, and the emission thus presents the 8:1 polarization ratio. Examples of Case (II) are the bleaching of the ground-state absorption due to ground-state depopulation, since it is the excitation *from* the nondegenerate orbital *to* the degenerate orbital (Case (II)-i), and the excited-state absorption in which the electron is excited from a lower orbital to the nondegenerate orbital (or in other words the hole in the nondegenerate orbital is excited to the "higher" orbital in terms of hole energy) (Case (II)-ii). Note that in this probe transition the initial "lower" orbital must be degenerate. In this case, the promoted electron does not carry the information of the pump polarization and the transition probability is isotropic in the molecular plane. Thus the probe projection factor for the parallel polarization configuration, $p_{\text{probe II}//}$, is

$$p_{\text{probe II}//} = \sin^2 \theta \tag{6}$$

and that for the perpendicular polarization configuration, $p_{\text{probe II}\perp}$, is

$$p_{\text{probe II}\perp} = (\cos \theta \cos \psi - \sin \psi)^2 \tag{7}$$

Hence, the average of the total projection factor for the random orientation is

$$\langle p_{\text{pump}} \, p_{\text{probe II}//} \rangle = \langle \sin^4 \theta \rangle = \frac{8}{15} \quad \text{Parallel configuration} \tag{8}$$

$$\langle p_{\text{pump}} \, p_{\text{probe II}\perp} \rangle = \langle \sin^2 \theta \, (\cos \theta \cos \psi - \sin \psi)^2 \rangle$$
$$= \frac{2}{5} \quad \text{Perpendicular configuration} \tag{9}$$

Therefore the parallel-to-perpendicular ratio is in this case 4:3. In Case (I), the anisotropy of the excited state in the plane might relax quickly so that there is no anisotropy in the plane, and the projection factor becomes the same as in Case (II) except for a factor of 1/2. Thus the parallel-to-perpendicular ratio is also 4:3 after in-plane relaxation.[6,7]

Anisotropy in another type of 2-D molecule, in which HOMOs are twofold degenerate and LUMO is nondegenerate, can be derived by simply exchanging electrons and holes in the picture shown in Figure 3. Case (I)-i is fluorescence, and case (I)-ii is excited-state absorption in which an electron in a lower orbital is promoted to the degenerate HOMO. These provide 8:1 polarization ratio. Case (II)-i is the bleaching due to ground-state depopulation, and case (II)-ii is the excited-state absorption in which the electron in the excited orbital is promoted to a higher orbital. These provide 4:3 ratio.

2.2. Experiment

The polarization dependence of the transient absorption spectrum was measured by means of a subpicosecond pump-probe spectroscopy in a solution of a vanadylphthalocyanine derivative. Due to its C_{4v} symmetry, the molecular orbitals have either nondegenerate a or b symmetry, or twofold degenerate e symmetry. Optical $\pi \rightarrow \pi^*$ transitions are polarized in the plane of the molecule and are symmetry-allowed only between nondegenerate (a or b) and degenerate (e) orbitals. This symmetry results into a twofold spatial degeneracy of $\pi \rightarrow \pi^*$ transitions. Indeed, the Q-band absorption, which is the sharp absorption band at about 700 nm, originates from the $a_{1u}(\pi) \rightarrow e_g(\pi^*)$ transition and is twofold degenerate in the molecular plane.[10-12] In order to avoid the influence of intermolecular interaction due to aggregation, a nonaggregative vanadylphthalocyanine derivative, hexadeca(trifluoroethoxy) vanadylphthalocyanine (VOPc(TFE)$_{16}$) shown in Figure 1,[19] was used. In the experiment, VOPc(TFE)$_{16}$ was dissolved in spectrograde ethyl acetate at a concentration of 5×10^{-4} mol/l, and optical measurements of the solution were made in a 1 mm thick quartz cell.

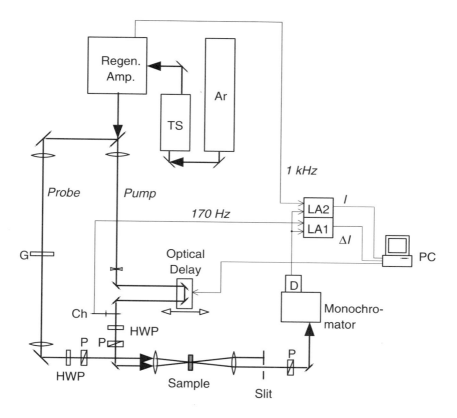

Figure 4 Schematic diagram of pump-probe experimental setup. Ar stands for Ar-ion laser, TS for mode-locked Ti-sapphire laser, G for glass plate for white continuum generation, Ch for optical chopper, P for polarizer, HWP for half-wave plate, D for photodiode, LA*i* for lock-in amplifier, and PC for personal computer.

The transient absorption change was measured using a typical sub-picosecond pump-probe measurement setup, as shown in Figure 4. The light source is an Ar^+ laser-pumped mode-locked Ti-sapphire laser with a Ti-sapphire regenerative amplifier, which provides about $10 \mu J$/pulse output with a temporal width of $250\,fs$ operating at $1\,kHz$. The probe beam is a spectral continuum of wavelengths from $450\,nm$ to longer than $1\,\mu m$ generated in a BK7 glass plate. Polarizer and half-wave plate combinations are used to control polarization and intensity of pump and probe, so that both the parallel and the perpendicular configurations are possible.

The transient absorption change of the $VOPc(TFE)_{16}$ solution shown in Figure 5 was approximately fitted by a biexponential function in the tested

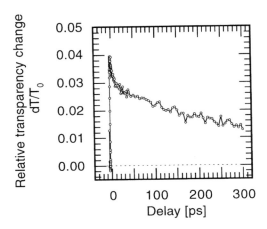

Figure 5 Time-resolved transmission change at a probe wavelength of 750 nm. Pump wavelength was 780 nm, and polarization of pump and probe was parallel.

region up to 300 ps delay, giving a short relaxation time of 4 ps and a long one of 400 ps in the parallel configuration and 4 ps and 1000 ps in the perpendicular configuration. The fast component is due to a spin-allowed fast intersystem crossing[20] originating in the spin degeneracy due to the unpaired electron in the vanadium d-orbital.[21] The difference in the slow relaxation time constant between the parallel and perpendicular configuration originates from depolarization due to the rotational diffusion of molecules in solution. The solution of the rotational diffusion equations is a multi-exponential function,[22] and the total response is a product of the rotational relaxation and the intrinsic electronic response. Assuming C_{4v} symmetry the rotational relaxation is expressed as

$$D_{//}(t) = \frac{2}{9} + \frac{2}{45}\exp\left(-\frac{t}{\tau_{\text{rot}}}\right) \quad \text{Parallel configuration} \quad (10)$$

$$D_{\perp}(t) = \frac{2}{9} - \frac{1}{45}\exp\left(-\frac{t}{\tau_{\text{rot}}}\right) \quad \text{Perpendicular configuration} \quad (11)$$

where τ_{rot} is the diffusion time constant of the rotation with an axis in the molecular plane. In this derivation the rotation with the axis normal to the molecular plane is not considered, assuming that there is no anisotropy of the nonlinear optical response in the plane as experimentally demonstrated later. The constant term is five or ten times larger than the exponentially decaying term, and thus $D_{//}(t)$ is a slightly decreasing function and $D_{\perp}(t)$ is a slightly increasing function. Therefore the total experimental response is approximately fitted by a biexponential function in the limited region of time delay, and in this fitting the slower component appears as if its decay time constant is modified negatively in the parallel configuration or positively in the perpendicular configuration. Fitting the experimental results including equations (10) and (11), a rotational diffusion time constant of 200 ps and the true lifetime of the species behind the slower component of 800 ps can be deduced.

Figures 6(a) and 6(b) show the transmission change spectra for parallel and perpendicular configurations. They were obtained by repeating the measurements varying the probe wavelength in 20 nm steps from 480 nm to 800 nm at delays of 1 ps (Figure 6(a)) and 20 ps (Figure 6(b)). The latter spectra represent the response of the long lived species: at this time delay the contribution of the fast component with the 4 ps time constant is negligible and the rotational diffusion has not yet affected the anisotropy of nonlinear optical response. On the other hand, the former spectra also include the contribution of the fast component. Comparing these spectral changes with the linear absorption spectrum, which is also shown in these figures, the bleaching of the Q-band is clearly observable as well as an extended excited-state absorption at the shorter wavelength, partially overlapping with each other at about 650 nm. In the spectra at 1 ps delay, a stimulated emission can be seen as shoulders at the fluorescence band located at the longer tail of the Q-band.

Figures 7(a) and 7(b) show the polarization ratio of the induced absorption change, i.e., the ratio of transient absorption change in the parallel configuration and in the perpendicular one, obtained from the results shown in Figures 6(a) and 6(b), respectively. The standard error of the measurements is also shown. At 1 ps delay, the ratio was approximately 4:3 except for a significant increase at the stimulated emission band. The stimulated emission behaves the same as the fluorescence to provide 8:1 ratio as the initial polarization dependence (Case I-(i)) and the ratio quickly decays to 4:3.[8] Therefore this result suggests that the in-plane electronic dephasing is not yet complete at 1 ps. Simultaneous appearance of 4:3 in the Q-band bleaching and higher ratio in the stimulated emission band clearly demonstrates the existence of an initial polarization ratio of 4:3 for the bleaching due to ground-state depopulation (Case II-(i)). At 20 ps delay the ratio was approximately 4:3 within the error over the whole range of wavelengths, both in the bleaching of the Q-band and in the excited-state absorption. Although the character of the transitions responsible for the excited-state absorption is not clarified yet, the 4:3 ratio at 1 ps delay suggests that the transition for the excited-state absorption possesses a character described as Case II-(ii).

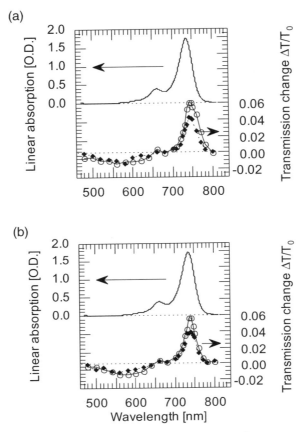

Figure 6   Top: linear absorption spectrum of the 5×10^{-4} mol/l VOPc(TFE)$_{16}$ ethyl acetate solution. Bottom: spectra of transmission change at (a) 1 ps and (b) 20 ps delay after excitation at 780 nm of the same VOPc(TFE)$_{16}$ solution. Open circles and solid line stand for the parallel polarization configuration, and closed diamonds and dashed line for the perpendicular polarization configuration. Note that this is a transmission change spectrum and that the positive sign corresponds to bleaching and the negative sign to induced absorption.

3.   POLING BEHAVIOR OF A DISUBSTITUTED CARBAZOLE: A NONRESONANT PROCESS IN A NONDEGENERATE 2-D SYSTEM

In this section, the anisotropy in refractive index of oriented disubstituted carbazoles, which have intramolecular CT transitions with 2-D character, is discussed. In such a nonresonant optical process, even nondegenerate

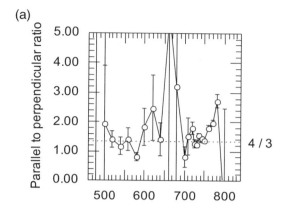

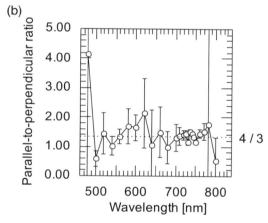

Figure 7   Ratio of induced absorption change between the parallel and perpendicular configuration of pump and probe polarization at (a) 1ps and (b) 20ps delay. Measurement errors are shown as vertical bars. The horizontal dashed line represents the ratio of 4/3.

transitions can contribute to its anisotropy. Carbazole derivatives can be model compounds for this investigation, since a carbazole substituted with two acceptors at 3- and 6-position presents 2-D intramolecular CT along the axes from N atom to the acceptors. This coupled CT results in a pair of CT transitions perpendicular to each other.[23] It has been shown that the optical properties and their poling behavior in a 2-D CT compound are quite different from those in a 1-D CT one.[23–27] It is expected that 2-D CT molecular design will provide larger off-diagonal components in nonlinearity and less

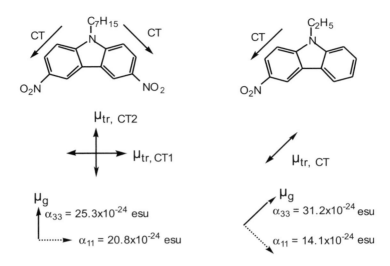

Figure 8 Top: molecular structure of 3,6-dinitro-N-heptylcarbazole (left) and 3-nitro-N-ethylcarbazole (right). Charge transfer axes are shown by arrows marked "CT". Middle: direction of CT transition dipole moments $\mu_{tr,CTi}$. Bottom: molecular polarizability components in the μ_g-coordinates for perfect uniaxial orientation. Direction of the ground-state dipole moment μ_g is also shown.

anisotropy in linear optical property. They are also interesting for the development of a "monolithic system" for photorefractivity.[25,26] Here we investigate 3,6-dinitro-N-heptylcarbazole with 2-D CT character and 3-nitro-N-ethylcarbazole with 1-D CT character (see Figure 8).

3.1. Theory

Molecular polarizability and macroscopic refractive index were obtained by semiempirical molecular orbital calculation.[23,24,28] In the calculation, 3,6-dinitro-N-ethylcarbazole was used in place of the N-heptyl derivative for simplicity. The molecular geometries were fully optimized at the Austin-Model 1 (AM1) level, and the intermediate-neglect-of-differential-overlap (INDO) single-configuration-interaction (SCI) technique was applied to obtain photoexcited states. The components of the molecular polarizability tensors were obtained by means of the sum-over-states (SOS) method.

The tensors were transformed into a Cartesian coordinates system in which the molecular ground-state dipole moment vector μ_g is z-axis (μ_g-coordinates), and they were averaged over rotation around z-axis. This corresponds to a perfect uniaxial orientation. The molecular polarizability

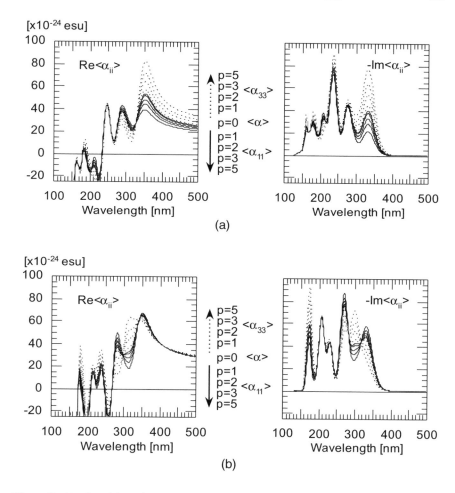

Figure 9 Real and imaginary parts of macroscopic molecular polarizability spectra at various poling factor p. (a) 1-D CT carbazole. (b) 2-D CT carbazole.

tensor components in the μ_g-coordinates at 1064 nm are $\alpha_{11} = 14.1$ and $\alpha_{33} = 31.2 \, [\times 10^{-24} \text{esu}]$ in 3-nitro-N-ethylcarbazole, and $\alpha_{11} = 21.1$ and $\alpha_{33} = 25.5 \, [\times 10^{-24} \text{esu}]$ in 3,6-dinitro-N-ethylcarbazole, as shown in Figure 8. In the 2-D CT compound, the molecular polarizability is less anisotropic than in the 1-D CT compound.

Macroscopic molecular polarizability tensors were obtained by transforming these μ_g-coordinates expressions into a Cartesian coordinates system in which the poling electric field E_{pol} is ζ-axis (macroscopic coordinates), and then by averaging over the direction of μ_g weighed by the

Boltzmann's factor. The linear polarizability tensor components are expressed as follows:

$$\langle \alpha_{11} \rangle = \alpha_{11} \frac{1 + L_2(p)}{2} + \alpha_{33} \frac{1 - L_2(p)}{2}, \tag{12}$$

$$\langle \alpha_{33} \rangle = \alpha_{11} (1 - L_2(p)) + \alpha_{33} L_2(p), \tag{13}$$

where $L_2(p)$ is the second order Langevin function, and p is the poling factor $|\mu_g||E_{\text{pol}}|/kT$. Figure 9 shows the macroscopic molecular polarizability spectra at various p value. Real and imaginary parts correspond to refractive index and absorption, respectively. It can be seen that the real part of 2-D CT carbazole presents very small change by orientation, due to its smaller anisotropy in the molecular polarizability in μ_g-coordinates. Changes in imaginary part by orientation present an opposite sign between 1-D and 2-D CT molecules, which is consistent with the theoretical prediction for perfect uniaxial orientation and the experimental result.[23]

Finally, macroscopic refractive indices n_i ($i = 1, 3$) were obtained from these macroscopic molecular polarizability tensor components by multiplying the number density of molecules and the local field factors. Figure 10 shows the evolution of the refractive index n_i of 10 wt% carbazole-doped poly(methyl methacrylate) (PMMA) with poling factor p. In the 2-D CT system, the birefringence induced by poling is much smaller than in the 1-D CT system, because of less anisotropy in the molecular polarizability. This is a great advantage of 2-D CT materials over 1-D CT materials in photonic device applications, since these devices are very sensitive to the refractive index of the material. Figure 11 shows the evolution of coherence length for SHG at 1064 nm in 10 wt% carbazole-doped PMMA. It should be noted that the scale for vertical axis is different between the 2-D and 1-D CT systems. The relative change in the 2-D CT material is nearly one order of magnitude smaller than the change in the 1-D CT material. Hence the 2-D CT material significantly improves the tolerance for orientation relaxation in photonic devices.

3.2. Experiment

Carbazole derivative was doped in PMMA with a concentration of 4.7 wt%, and spin-coated on a Pyrex glass substrate. The sample was dried in vacuum at 100 °C for several hours, and was divided into two. After measurement of absorption spectrum, these samples were simultaneously corona-poled at 15 kV for one minute at 105 °C. Anisotropic refractive indices were measured by the m-line (mode line) method at 632.8 nm with one sample, and the absorption spectrum was measured at the same time with the other

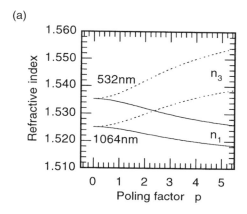

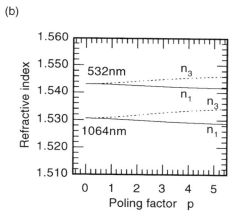

Figure 10 Evolution of refractive index n_1 and n_3 with poling factor p. Full lines represent the refractive indices for TE polarization n_1, and dashed lines represent that for TM polarization n_3. (a) 1-D CT carbazole. (b) 2-D CT carbazole.

sample. After certain period of measurement (about 100 hours), the samples were heated at 105 °C for complete orientation relaxation, and the refractive index and absorption were measured to confirm no sample degradation.

Figure 12 shows the evolution of order parameter obtained from the absorption change. Order parameter θ and relative absorption change $\Delta = A/A_0 - 1$ (where A and A_0 are absorbance at λ_{max} after and before poling) are related as:

$$\theta = -\Delta, \quad \text{1-D CT molecule} \tag{14}$$

$$\theta = 2\Delta. \quad \text{2-D CT molecule} \tag{15}$$

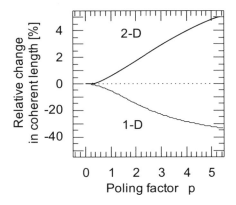

Figure 11 Evolution of coherence length for SHG in 10 wt% carbazole-doped PMMA with poling factor p. Fundamental wavelength is 1064 nm, and TE polarization is assumed in fundamental and SH field. Coherent length at $p = 0$ is 21 μm for 2-D CT dinitrocarbazole, and 26 μm for 1-D CT nitrocarbazole.

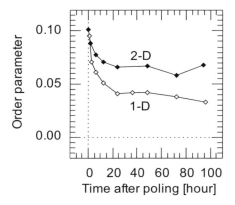

Figure 12 Order parameters and their relaxation estimated from absorption change.

The order parameter just after poling was about 0.1 and not so different between the 1-D and 2-D CT carbazoles. This is consistent with the fact that the ground-state dipole moments of the 1-D and 2-D CT carbazoles are almost equal. After initial relaxation for about 24 hours, the order parameter reduced to 0.07 in the 2-D CT sample and 0.04 in the 1-D CT one, followed by slow relaxation.

(a)

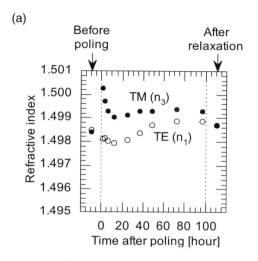

(b)

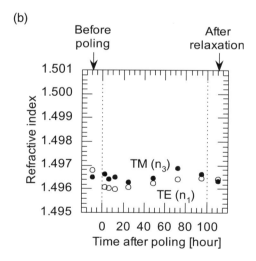

Figure 13   Anisotropic refractive indices measured by the m-line method. (a) 1-D CT carbazole. (b) 2-D CT carbazole. Refractive indices measured before poling and after complete orientation relaxation are also shown.

Figure 13 shows the poling behavior of refractive indices measured by the m-line method. The 2-D CT carbazole presented only a small birefringence of 0.0006 at maximum, which did not change significantly during the whole measurement. On the contrary, the 1-D CT carbazole presented a large initial birefringence as large as 0.0022 which significantly reduced to 0.0011 after 24 hours and to 0.0005 after 96 hours, which was still larger than the value of 2-D CT carbazole even though the order parameter was smaller. Refractive indices before poling and after complete relaxation were coincide within measurement error (estimated as 0.0003). This experimental result is consistent with the theoretical prediction, demonstrating the advantage of a 2-D CT material in photonic device applications that it is more tolerant to orientational relaxation.

4.   CONCLUSION

Anisotropy in the linear and nonlinear optical properties of a molecular system has been discussed in terms of the dimensionality of the component molecule. Two categories of optical response in 2-D molecules were invest-igated, and it was shown that they present an anisotropy different from ordinary 1-D molecules.

The first one was the anisotropy of transient photoinduced absorption change in a metallophthalocyanine with C4v symmetry. It has been derived that the polarization dependence in pump-probe response is 4:3 (less ani-sotropic than a 1-D molecule) or 8:1 (more anisotropic), depending on the nature of the transition for probe whether it carries the information of pump polarization or not. Absorption saturation due to the ground-state depopulation is an example for the former case to present 4:3 ratio, and fluorescence is an example of the latter case to present 8:1 ratio. Sub-picosecond transient absorption spectra were measured by the pump-probe method using a nonaggregative vanadylphthalocyanine solution, and the ratio of 4:3 was experimentally demonstrated. Depolarization due to the molecular orientation relaxation was analyzed for 2-D molecule, and a specific depolarization function due to two-dimensionality was derived.

The second one was the anisotropy of refractive index in a poled polymer based on a disubstituted carbazole with 2-D CT character. It was theore-tically predicted that the molecular polarizability is much more isotropic than its 1-D CT counterpart, resulting in smaller anisotropy and smaller change in refractive index due to orientation relaxation. Advantage of this feature in photonic device applications is discussed in terms of tolerance to orientation relaxation, and as an example the poling behavior of the coherence length for SHG was theoretically evaluated. In experiment, ani-sotropic refractive indices of PMMA films doped with 2-D and 1-D CT carbazoles were measured by the m-line method. Anisotropy of refractive index by poling and its change due to orientation relaxation was sig-nificantly smaller in the 2-D CT carbazole, demonstrating the advantage of 2-D CT molecules in photonic device applications.

ACKNOWLEDGEMENTS

The authors expresses their gratitude to Dr. Nicola Pfeffer, Dr. Yadong Zhang, Dr. Minquan Tian, and Ms. Mariko Tsuyuki for their collaboration. One of the author (T.I.) expresses his acknowledge to the Institute of Physical and Chemical Research (RIKEN) and Science and Technology Agency of Japan (STA) for supporting him through the "Special Post-doctoral Researcher for Basic Science" Program of RIKEN.

REFERENCES

1. Newkome, G.R., Moorefield, C.N. and Vögtle, F. (1996). *Dendritic Molecules: concepts, syntheses, perspectives* (VCH, Weinheim).
2. Ho, P.P. and Alfano, R.R. (1979). *Phys. Rev. A*, **20**, 2170–2187.
3. Moerner, W.E., Silence, S.M., Hache, F. and Bjorklund, G.C. (1994). *J. Opt. Soc. Am. B*, **11**, 320–330.
4. Perrin, F. (1929). *Ann. Phys.*, (Paris) **12**, 169–275.
5. Gouterman, M. and Stryer, L. (1962). *J. Chem. Phys.*, **37**, 2260–2266.
6. Knox, R.S. and Gilmore, R.C. (1995). *J. Luminescence*, **63**, 163–175.
7. Wynne, K. and Hochstrasser, R.M. (1993). *Chem. Phys.*, **171**, 179–188.
8. Galli, C., Wynne, K., LeCours, S.M., Therien, M.J. and Hochstrasser, R.M. (1993). *Chem. Phys. Lett.*, **206**, 493–499.
9. Pfeffer, N., Isoshima, T., Tian, M., Wada, T., Nunzi, J.-M. and Sasabe, H. (1997). *Phys. Rev. A*, **55**, R2507–R2510.
10. Leznoff, C.C. and Lever, A.B.P. (1989). *Phthalocyanines: properties and applications* (VCH Publishers, New York).
11. Lee, L.K., Sabelli, N.H. and LeBreton, P.R. (1982). *J. Phys. Chem.*, **86**, 3926–3931.
12. Orti, E., Brédas, J.L. and Clarisse, C. (1990). *J. Chem. Phys.*, **92**, 1228–1235.
13. Thomas, A.L. (1990). *Phthalocyanine: research and applications* (CRC Press, Boca Raton).
14. Perry, J.W., Mansour, K., Marder, S.R., Perry, K.J., Daniel Alvarez, J. and Choong, I. (1994). *Opt. Lett.*, **19**, 625–627.
15. Li, C., Zhang, L., Yang, M., Wang, H. and Wang, Y. (1994). *Phys. Rev. A*, **49**, 1149–1157.
16. Zhou, Q.L., Heflin, J.R., Wong, K.Y. Zamani-Khamiri, O. and Garito, A.F. (1991). *Phys. Rev. A*, **43**, 1673–1676.
17. Rodenberger, D.C., Heflin, J.R. and Garito, A.F. (1992). *Nature* (London) **359**, 309–311.
18. Isoshima, T., Pfeffer, N., Tian, M., Wada, T. and Sasabe, H. (1996). *Proc. Photo. Opt. Soc.*, (SPIE) **2852**, 211–219.
19. Wada, T. and Sasabe, H. (1994). *Proc. Photo. Opt. Soc.*, (SPIE) **2143**, 164–171.
20. Terasaki, A., Hosoda, M., Wada, T., Tada, H., Koma, A., Yamada, A., Sasabe, H., Garito, A.F. and Kobayashi, T. (1992). *J. Phys. Chem.*, **96**, 10534–10542.
21. Ake, R.L. and Gouterman, M. (1969). *Theoret. Chim. Acta*, **15**, 20–42.
22. Favro, L.D. (1960). *Phys. Rev.*, **119**, 53–62.
23. Isoshima, T., Wada, T., Zhang, Y.-D., Brouyère, E., Brédas, J.-L. and Sasabe, H. (1996). *J. Chem. Phys.*, **104**, 2467–2475.
24. Isoshima, T., Zhang, Y.-D., Brouyère, E., Brédas, J.-L., Wada, T. and Sasabe, H. (1995). *Nonlinear Opt.*, **14**, 175–183.
25. Wada, T., Zhang, Y., Choi, Y.S. and Sasabe, H. (1993). *J. Phys. D: Appl. Phys.*, **26**, B221–B224.

26. Wada, T., Zhang, Y., Yamakado, M. and Sasabe, H. (1993). *Mol. Cryst. Liq. Cryst.*, **227**, 85–92.
27. Watanabe, T., Kagami, M., Miyamoto, H., Kidoguchi, A. and Miyata, S. (1992). In *Nonlinear Optics. Fundamentals, Materials and Devices. Proceedings of the Fifth Toyota Conference on Nonlinear Optical Materials*, edited by S. Miyata (Elsevier Science Publishers, Aichi, Japan), pp. 201–206.
28. Isoshima, T., Wada, T., Zhang, Y.-D., Aoyama, T., Brédas, J.-L. and Sasabe, H. (1996). *Nonlinear Opt.*, **15**, 65–68.

11. MESOSCOPIC PATTERN FORMATION OF NANOSTRUCTURED POLYMER ASSEMBLIES: HIERARCHICAL STRUCTURING OF TWO-DIMENSIONAL MOLECULAR ORGANIZATES

MASATSUGU SHIMOMURA*, NORIHIKO MARUYAMA,
TAKEO KOITO and OLAF KARTHAUS

*Research Institute for Electronic Science, Hokkaido University,
Sapporo 060-0812, Japan*

INTRODUCTION

A large interest is increasing in a biomimetic approach for materials design (Aksay, 1996). Hierarchical structuring of molecules is vital and constitutional base of living organs. Biological membrane, which is a typical nanometer-size self-assembly of biological molecules, is an essential structural component of mesoscopic subcellular apparatus, i.e., organellas. A biological cell is a micrometer-size assembly of the organellas and a constructing unit of biological tissues and living organs in macroscopic scale. Hierarchical structuring is an indispensable strategy of the biomimetic approach for molecular architectures of the hyper-structured molecular devices. In the last decade, by using the self-assembling nature of artificially designed molecules chemists have succeeded to construct many sorts of nanometer-size molecular assemblies, *e.g.*, molecular-recognition-directed molecular assemblies (Lehn, 1990), surfactant bilayer assemblies (Kunitake, 1992), and polymeric Langmuir-Blodgett films (Ringsdorf, 1988).

The driving forces of molecular assembling employed in the nanoscopic world are many week physico-chemical intermolecular interactions, *e.g.*, van der Waals force, hydrogen bonding, hydrophobic interaction, electrostatic force, and so on. Hierarchical structuring of the nanometer-size molecular assemblies up to the mesoscopic region, from submicrometer to submilimeter, requires another guiding principle of self-organization. As a general physical phenomenon, we know another type of self-organization, so-called "dissipative structures" which are generated far from thermal equilibrium. Many dynamic structures of self-organization with various spatial scales,

* Tel.: +81-11-706-2997, Fax: +81-11-706-4974, E-mail: shimo@poly.es.hokudai.ac.jp

e.g., Rayleigh-Bénard convection in heated liquids, spirals in the Belousov-Zhabotinsky reaction, are known as dissipative structures.

Since the dissipative structure is essentially a dynamic structure with continual energy dissipation, the structures have to be frozen as statically stable structures of molecular assemblies. We focused a casting process of polymer solution on a solid surface because the casting processes affected by many physical parameters, *e.g.*, viscosity, temperature, surface tension, etc. are complex enough to form the dissipative structures. And after rapid solvent evaporation, the dissipative structures formed in the casting solution can be immobilized as the regular polymer patterns on solid surface. Here we report a novel general method of the hierarchical structuring of molecular assemblies based on freezing the spatiotemporal structures formed in a casting process of polymer solutions.

BILAYER MEMBRANES AS NANOSTRUCTURED MOLECULAR ASSEMBLIES

Bilayer membranes are two-dimensional molecular aggregates spontaneously assembled in water and are representatively known as biomembranes and liposomes in biological science. Since Kunitake found the formation of the bilayer membrane from a simple double-chain ammonium salt **1** in 1977 (Kunitake, 1977), a large variety of synthetic amphiphiles which are not directly related to the structure of biolipids has been reported as bilayer-forming molecules. Single-chain surfactants having aromatic π-electron groups can form the bilayer membranes, too (Shimomura, 1993). One of the unique structural characteristics of the bilayer membrane is two-dimensional molecular ordering with regularly molecular packing. Oriented π-electron arrays are expected to act as electronic media for photoexcited electron and/or energy transfer processes. Bilayer membranes of the single-chain amphiphiles are suitable candidates for the hyper-structured molecular devices with photonic functions based on their two-dimensional molecular ordering.

One of the representative photonic functions of the stacked π-electron arrays in organic materials is efficient energy migration. Photoexcitation energy is delocalized as a "molecular exciton" (Frenkel exciton) in molecular crystals having the highly ordered arrangement of π-electrons. An efficient energy transfer based on Frenkel exciton in the two-dimensional molecular assemblies is expected as well as in organic molecular crystals. Intermolecular interaction is classified by three categories depending on the intensity of the intermolecular coupling force. In the case of strong interaction, intermolecular energy transfer is faster than intramolecular vibrational relaxation from Frank-Condon state. A strong intermolecular interaction at

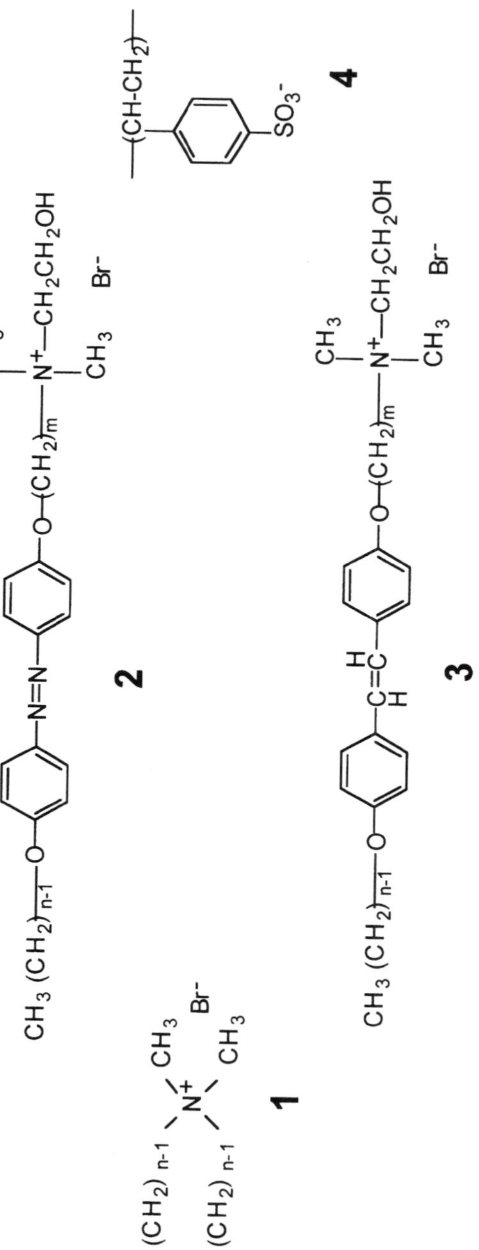

Scheme 1

the ground state is observed as Davydov splitting in the absorption spectrum. The rate of intermolecular energy transfer is comparable to the vibrational relaxation in weak-coupling interaction. Spectral shift is not observed, but change of molar absorption coefficient such as hypochromism or hyperchromism is often found. While in the very weak-coupling case, excitation energy transfers after the vibrational relaxation. Resonance transfer mechanism based on the dipole-dipole interaction proposed by Förster is applicable to the very weak interaction. Kasha proposed "molecular exciton theory" for the strong-coupling case in molecular aggregates (Kasha, 1976).

A single-chain azobenzene amphiphile **2** whose absorption spectrum is strongly affected by the intermolecular interaction in the bilayer membrane (Shimomura, 1983). The absorption maximum of the azobenzene chromophore is located at 355 nm when the amphiphile is molecularly dispersed in ethanol. While in the bilayer membrane in water, an extensive spectral variation is observed for the identical chromophore. Absorption maximum is strongly dependent on the alkyl chain length. Homologous series of $m = 5$ show a large bathochromic shift to around 400 nm. Hypsochromic shift to 300 nm is found when the difference of the two alkyl chain (m-n) is larger than two (Figure 1(a)). The Davydov splitting in the absorption spectrum of the azobenzene bilayer membranes strongly suggests the formation of Frenkel excitons which are attributed to a strong intermolecular coupling among the excited states of the aromatic π-electron groups. According to the semi-quantitative calculation of molecular exciton theory of Kasha, the blue and red shift relative to the isolated azobenzene chromophore are attributed to side-by-side and head-to-tail orientation in the bilayer membrane, respectively (Figure 1(b)). An X-ray structural analysis of the single crystal proves that the spectral estimation on the molecular packing is in fact (Okuyama, 1993).

A broad emission with a large Stokes shift is observed the azobenzene amphiphile ($m = 5$) forms the bilayer membrane with the head-to-tail chromophore orientation (Shimomura, 1987). Time-resolved fluorescence spectra are shown in Figure 2. At the early stage of the photoexcitation, a sharp emission with very short life time (*ca.* 10 picosecond) is found at around 470 nm. A new emission at around 600 nm appears concomitant with the decay of 470 nm emission. Because the excitation spectrum monitored at 600 nm emission is identical to the absorption spectrum, the peculiar emission of the azobenzene bilayer is ascribable to the exciton emission. The short lived emission is attributed to a free exciton that migrates around in the bilayer membrane and is captured so-called "trap" site (self-trapped exciton).

Absorption spectrum of a stilbene amphiphile **3** ($n = 12$, $m = 10$) in ethanol shows a typical spectral pattern of trans isomer (dashed line in Figure 3(a)), whereas a large hypsochromic shift was observed in an

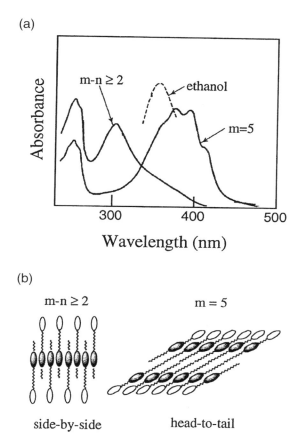

Figure 1 Absorption spectra (a) and molecular orientation models (b) of the azobenzene bilayer membranes

aqueous bilayer solution. This shift was apparently attributed to the parallel chromophore orientation shown in Figure 1(b). Red-shifted structured fluorescence was ascribable to the emission from the exciton state. As shown in Figure 3(a), fluorescence of the stilbene exciton was effectively quenched by fluorescein as an anionic energy acceptor and the emission of the acceptor appeared at 530 nm. A straight line of Stern-Volmer plot (I_0/I vs. the acceptor concentration, where I_0 and I are the fluorescence intensity of the donor in the absence and presence of the acceptor molecule, respectively) indicates that the quenching process is diffusion controlled. The slope of the plot gives Stern-Volmer constant that is the product of the donor's life time and the quenching rate constant. A large value (1.8×10^7) of the

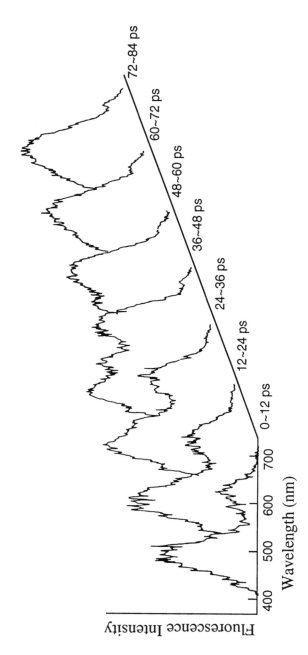

Figure 2   Time-resolved fluorescence spectra of the azobenzene bilayer membrane ($m = 5$, $n = 12$) after 315 nm excitation.

(a)

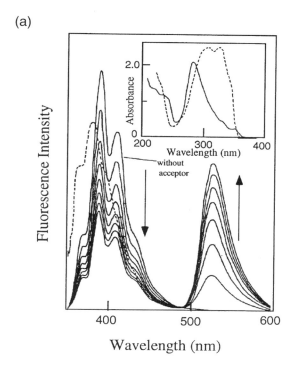

(b)

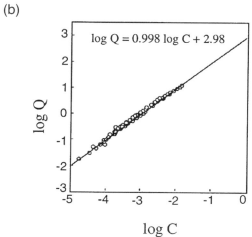

Figure 3 Absorption spectra and fluorescence quenching of the stilbene bilayer membrane (a) and Klöpffer's plots of energy transfer (b). Broken curves in (a) show spectrum of an ethanol solution.

Stern-Volmer constant of this experiment suggests that the quenching rate is extremely faster than the diffusion rate of the acceptor molecule even if the life time of the donor is estimated to be one nanosecond. Klöpffer proposed quenching kinetics based on the exciton diffusion for organic molecular crystals (Klöpffer, 1969). In this mechanism, quenching efficiency Q ($Q = I_0/I - 1$) is proportional to the trapping probability, which is given by a product among the jumping probability during the life time, the capturing probability by the acceptor site, and the molar ratio of the acceptor molecule. If the capturing probability is unity and only acceptor site acts as energy trap (no self-trapped exciton), the quenching kinetics is expressed in a simple equation

$$Q = nc \, (1 - F) = n'c, \tag{1}$$

where c is the acceptor concentration (mol/mol), n is the hopping number of the exciton, F is the returning probability to the starting point in the random walk process of the exciton hopping. A straight line with the slope of 1.0 and the intercept ($\log n'$) of 3.0 in the double-logarithmic plots of equation (1), shown in Figure 3(b), strongly indicates that the stilbene exciton hops around the bilayer membrane more than 1000 times within its lifetime. Effective photoenergy collection based on the singlet exciton hopping in the stilbene bilayer assemblies strongly suggests that two-dimensional π-electron arrays are indispensable in designing hyper-structured molecular devices having photonic functions.

POLYION COMPLEXES AS NANOSTRUCTURED POLYMER ASSEMBLIES

Immobilization of the bilayer membranes as thin solid films is required when the bilayer membranes are used as hyper-structured molecular devices. Casting method is a simple way to immobilize the bilayer membrane on a solid support from an aqueous solution by drying process. A cast film was used as a sample of the X-ray structural experiment since the layered structure of the bilayer membrane were easily observed (Kunitake, 1984). The cast film of the azobenzene amphiphile is easily peeled off as a self-standing film from the solid substrate. Polymerization of the bilayer films is another way of materials fabrication. Polyion complex technique is a unique method for immobilization of bilayer membranes with polymers. Water-insoluble complex is precipitated as the polyion complex when the aqueous solution of the charged bilayer membrane is mixed with a water solution of the counter charged polyelectrolyte. Polyion complexes are the nanoscopic molecular architectures with regularly ordered two-dimensional layered structures at the air-water interface (Shimomura, 1985) and in the solvent cast films (Shimomura, 1991). The fundamental bilayer structure is mostly maintained in the immobilized film. X-ray diffraction of the solvent cast films indicates the layered structure with repeating spacing corresponding to the bilayer thickness (Okuyama, 1996). Two-dimensional molecular ordering with regularly molecular packing in the bilayer membranes is also immobilized in the polyion complex films. Absorption maximum of the cast film of the azobenzene amphiphiles complexed with anionic polymers, *e.g.*, poly(styrene sulfonate) **4**, is almost identical to that of the original bilayer membranes without polymer.

MESOSCOPIC STRUCTURING OF POLYION COMPLEXES

Since the hierarchical structuring in the mesoscopic range, *e.g.*, micro phase separation in block copolymers, is considerably indispensable for designing new polymer materials, we attempt to use the dissipative processes for the mesoscopic structuring of the nanostructured polymer assemblies composed from polyelectrolytes and bilayer-forming amphiphiles. A water-insoluble polyion complex was precipitated when an aqueous bilayer solution of **1** was mixed with a water solution of **4**. Several microliters of the highly diluted polymer solution (a few hundred micrograms per liter) was spread homogeneously as a liquid film over *ca.* 1 cm^2 area on clean hydrophilic surface of glass, silicon wafer, and freshly cleaved mica in a glove box under controlled atmospheric humidity. To visualize the evaporation process through an epifluorescence microscope a small amount of an amphiphilic cationic fluorescence probe, octadecyl rhodamin B, was added as a counter ion of the polyanion as well as the bilayer-forming amphiphiles.

Figure 4(a) is an instantaneous top view of the casting solution. Typical dissipative structures of Bénard-convection cells were formed as several circular domains in the central portion of the polymer solution. Bright streams from the cell boundary toward the solution front indicate another convectional flow between the three-phase line (liquid-substrate-air boundary) and the cell boundary. Condensation of polymer at the three-phase line can be observed directly by fluorescence microscopy. The solution front was clearly observed as a bright red boundary line between the dark substrate and the bulk solution. An interesting finding was a periodic polymer condensation along the three-phase line. The periodic fingering instability at the three-phase line is assumed to be generated by the radial convectional streams originated from the regular arrangement of the Bénard-type convection cells in the central portion. On the surface of the freshly cleaved mica, the fingers often started to grow up perpendicular to the three-phase line. Figure 4(b) shows snap shots of the stripe formation from the periodically generated fingers. The fingers were straightened as

(a)   (b)

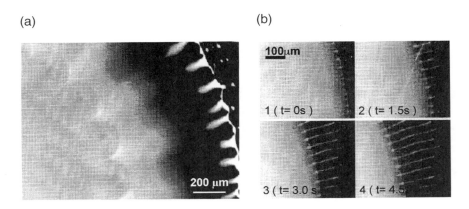

Figure 4 *In situ* epifluorescence microscopic observations of the casting solution (top view) during solvent evaporation (a) and snap shots of stripe formation from the three-phase line (b).

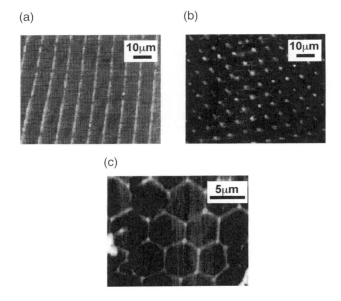

Figure 5 Atomic force micrographs of polymer patterns fixed on solid surface after solvent evaporation. (a) stripes, (b) regularly arranged dots, (c) honeycomb structure.

regular stripes concomitant with smooth receding of the solution front. Figure 5(a) is an atomic force micrograph of the stripe pattern formed on mica surface. Dimensions of the stripe patterns are altered by casting conditions. The higher concentration and molecular weight of polymers brings the larger dimensions of the stripe. A regularly arranged dot pattern (Figure 5(b)) was occasionally formed when the dewetting of polymer was occurred in each stripe.

At the final stage of the casting, homogeneous and optically transparent polymer films were prepared under the dry casting condition. Under high atmospheric humidity, however, the films were not optically transparent. Figure 5(c) shows an atomic force image of the cast film prepared under the wet condition. The cast film has regular honeycomb morphology with a size of a few micrometer per each honeycomb cell. Based on the *in-situ* observation of the casting process, the formation mechanism of the honeycomb structure is schematically shown in Figure 6. After placing a droplet of chloroform solution on the substrate, the chloroform starts to evaporate. This leads to a cooling of the solution and micron-size water droplets condense onto the chloroform solution of the polyion complex. The droplets are transported to the three-phase line and are hexagonally packed by the convectional flow or the capillary force generated at the solution front. Since the surface tension between water and chloroform is reduced by the polyion complex of the bilayer forming surfactants, the water droplets are stabilized against fusion. Upon evaporation of chloroform the three-phase line moves over the hexagonal array of water droplets. The water droplets and some of the polymer between them are left behind. Finally the water evaporates, leading to the observed honeycomb structure. Regularly ordered honeycomb patterns of polymer assemblies formed by a similar mechanism were proposed by Widawski *et al.* (Widawski, 1994).

The most important parameters to control the size of the honeycomb cells are the relative humidity of the atmosphere and the concentration of the polymer solution. The honeycomb wall surrounding the hole becomes thinner with decreasing concentration, and higher humidity leads the honeycomb holes to be larger. The size of the honeycomb holes can be regulated between 0.5 and 5 micrometers by the two parameters. The X-ray diffraction pattern of the honeycomb film with Bragg peaks of 3.6 nm repeating period strongly suggests that the honeycomb walls are constituted of nanostructured lamella assemblies of the polyion complexes. UV-visible absorption spectrum of the honeycomb film prepared from the azobenzene amphiphile is identical to that of the bilayer membranes.

CONCLUSION

It is concluded that the spatiotemporal structures based on dissipative structures in polymer solutions are freezed on solid surfaces as the regular

(a)

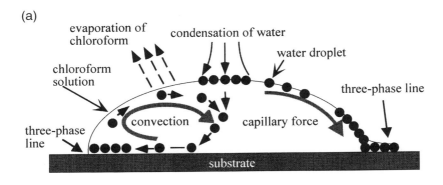

(b)

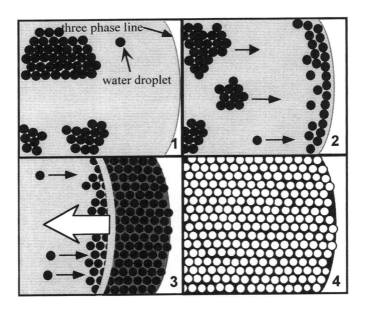

Figure 6   Schematic formation mechanism of the honeycomb structure, from side view (a) and top view (b).

mesoscopic polymer assemblies with the hierarchical structures of the nanoscopic bilayer assemblies (Figure 7). We believe that our finding is applicable as a novel general method of the mesoscopic structure formation without lithographic procedures because the formation of dissipative

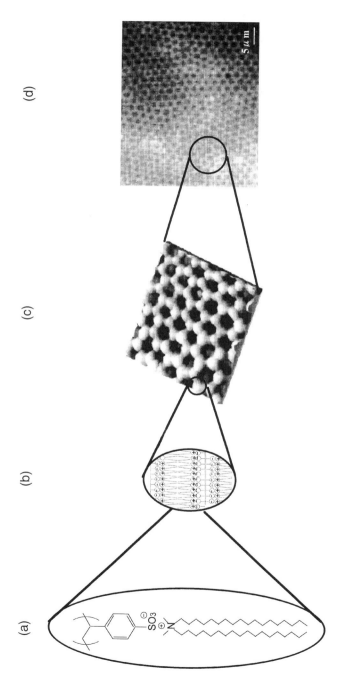

Figure 7 Schematic illustration of the hierarchical structuring of nanostructured polymer assemblies. (a) molecular structure, (b) nanostructured bilayer assemblies, (c) mesoscopic pattern in the range of the atomic force microscopy, (d) micrometer scale structure.

structure is essentially a general physical phenomenon for any polymer materials. To fabricate the mesoscopic structure with good reproducibility, we are now preparing the regular stripes under the controlled continuous front receding, e.g., generated in the dipping up procedure of solid substrates from a polymer solution in the controlled conditions (dipping speed, temperature, solvent evaporation, etc.). We have already prepared long stripe structures of double-helical DNA complexed with 1 on mica surface to measure the photoinduced electron transfer through the highly oriented π-electron arrays in the stacked base-pairings (Shimomura, 1997).

REFERENCES

Aksay, I.A., Trau, M., Manne, S., Honma, I., Yao, N., Zhou, L., Fenter, P., Eisenberger, P.M. and Gruner, S.M. (1996). Biomimetic Pathways for Assembling Inorganic Thin Films. *Science*, **273**, 892.

Lehn, J.M. (1990). Perspectives in Supramolecular Chemistry, *Angew. Chem. Int. Ed. Engl.*, **29**, 1304.

Kunitake, T. (1992). Synthetic Bilayer Membranes: Molecular Design, Self-Organization, and Application. *Angew. Chem., Int. Ed. Engle.*, **31**, 709.

Ringsdorf, H., Schlarb, B. and Venzmer, J. (1988). Molecular Architecture and Function of Polymeric Oriented Systems: Models for the Study of Organization, Surface Recognition, and Dynamics of Biomembranes. *Angew. Chem. Int. Ed. Engl.*, **27**, 139.

Kunitake, T. and Okahata, Y. (1977). A Totally Synthetic Bilayer Membrane. *J. Am. Chem. Soc.*, **99**, 3860.

Shimomura, M. (1993). Preparation of Ultrathin Polymer Films Based on Two-Dimensional Molecular Ordering. *Prog. Polym. Sci.*, **18**, 295.

Kasha, M. (1976). Molecular Exciton in Small Aggregates. *Spectroscopy of the Excited State*, 337 (ed. by Bartolo, B.D., Plenum Press).

Shimomura, M., Ando, R. and Kunitake, T. (1983). Orientation and Spectral Characteristics of the Azobenzene Chromophores in the Ammonium Bilayer Assembly. *Ber. Bunsenges. Phys. Chem.*, **87**, 1134–1143.

Okuyama, K. and Shimomura, M. (1993). Structural Diversity of Bilayer-Forming Synthetic Lipids. *New Functionality Materials, Volume C, Synthetic Process and Control of Functionality Materials*, 503–508 (ed. by Tsuruta, T., Doyama, M. and Seno, M. Elsevier Science Publishers B.V.).

Shimomura, M. and Kunitake, T. (1987). Fluorescence and Photoisomerization of Azobenzene-Containing Bilayer Membranes. *J. Am. Chem. Soc.*, **109**, 5175–5183.

Shimomura, M., Hashimoto, H. and Kunitake, T. (1989). Controlled Stilbene Photochemistry in Ammonium Bilayer Membranes. *Langmuir*, **5**, 174–180.

Klöpffer, K.J. (1969). Transfer of Electronic Excitation Energy in Solid Solutions of Perylene in N-Isopropylcarbazole. *J. Chem. Phys.*, **50**, 1689.

Kunitake, T., Shimomura, M., Harada, A., Okuyama, K., Kajiyama, T. and Takayanagi, M. (1984). Ordered Cast Film of an Azobenzene-containing Molecular Membrane. *Thin Solid Films*, **121**, L89-L91.

Shimomura, M. and Kunitake, T. (1985). Preparation of Langmuir-Blodgett Films of Azobenzene Amphiphiles as Polyion Complexes. *Thin Solid Films*. **132**, 243–248.

Shimomura, M., Utsugi, K., Horikoshi, J., Okuyama, K., Hatozaki, O. and Oyama, N. (1991). Two-Dimensional Ordering of Viologen Polymers Fixed on Charged Surface of Bilayer

Membranes: A Peculiar Odd-Even Effect on Redox Potential and Absorption Spectrum. *Langmuir*, **7**, 760–765.

Okuyama, K. and Shimomura, M. (1996). Crystal Engineering of Synthetic Bilayer Membranes *New Developments in Construction and Functions of Organic Thin Films*, 39–70 (ed. by Kajiyama T. and Aizawa, M. Elsevier Science B.V.).

Widawski, G., Rawiso, M. and François, B. (1994). Self-organized Honeycomb Morphology of Star-polymer Polystyrene Films. *Nature*, **369**, 387.

Shimomura, M., Karthaus, O., Maruyama, N., Ijiro, K., Sawadaishi, T., Tokura, S. and Nishi, N. (1997). Mesoscopic Patterns of DNA-Amphiphile Complexes. *Rep. Prog. Polym. Phys. Jpn.*, **40**, 523–524.

12. SELF-ASSEMBLY OF NANOSTRUCTURED MATERIALS: DREAMS AND REALITY

DONALD BETHELL, DAVID J. SCHIFFRIN,
CHRISTOPHER KIELY, MATHIAS BRUST and JOHN FINK

*Department of Chemistry and Department of Materials Science and
Engineering, University of Liverpool, Liverpool L69 3BX, UK*

Research in this field in Liverpool began three or four years ago with a brief conversation between two of the present authors which went as follows:

Schiffrin: "What you organic chemists should be doing is synthesising a brain."
Bethell: "Yes, but can we start with something simpler?"

Out of that brief exchange grew the idea that it might be possible to construct arrays of very small metallic particles within a self-assembled organic matrix and that electron transport between the particles might be controlled by taking advantage of certain well-known properties of small metal clusters which possess a distribution of electronic energy levels that lies intermediate between the band structure of bulk metals and the widely spaced levels to be found for example in simple organic molecules (Schmid, 1992). The plan of attack on the problem is illustrated in Figure 1. Initially, nanometre-sized metal particles, particularly gold particles were to be generated, and a layer of alkanethiol deposited on the surface using self-assembly which was already well-established on planar gold surfaces. In Stage 2, again using self-assembly, methods would be sought to achieve a more ordered array of gold particles in a continuous organic matrix. Finally, by incorporating externally addressable electro- or photo-active groups into the organic matrix positioned between adjacent metal particles, it was hoped that electron transport could be modulated. This, then, was the original dream mentioned in the title. Such materials, it is hoped, can be regarded as being close to the hyperstructured organic molecules that are the subject of this Forum. What will be described is the work that has been carried out in Liverpool; it should be stated at the outset that we were so enthusiastic about the potential of the materials produced in our early experiments that we sought to develop our investigations in several different directions simultaneously. As a result a number of fascinating observations will be described for which we can offer only partial interpretations (or no explanation at all) because we have not had the manpower and facilities to undertake the necessary detailed investigation. Elegant studies are now appearing in the literature, most notably

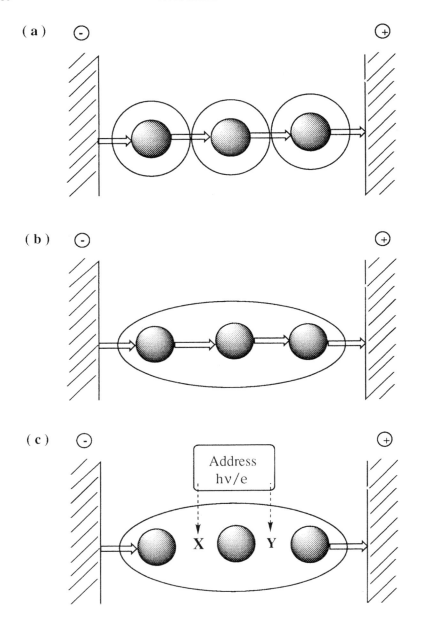

Figure 1 Proposed development of nanostructured materials based on colloidal gold particles: (a) assemblies of individually thiol-capped gold nanoparticles; (b) gold nanoparticles in a continuous self-assembled organic matrix; (c) gold nanoparticles in a matrix incorporating addressable organic functionalities X and Y.

from the laboratories of Murray (Terrill *et al.*, 1995; Hostetler *et al.*, 1996); Whetten (Whetten *et al.*, 1996); Lennox (Badia *et al.*, 1996, 1997) and others, on nanostructured materials often prepared using the methods that we have developed, and our current thinking owes much to their careful, painstaking and laborious research.

PREPARATION AND CHARACTERISATION OF GOLD NANOPARTICLES

Our preparative procedure is remarkably simple, combining well established principles of colloid chemistry, phase transfer between immiscible liquids and self assembly on solid surfaces (Brust *et al.*, 1994). A solution in water of a tetrachloraurate salt is added to toluene and a phase-transfer agent, typically tetraoctylammonim bromide is added. The initially pale yellow aqueous layer becomes colourless while the upper toluene layer takes on a deep orange hue. Addition of an aqueous solution of sodium borohydride with vigorous stirring then leads to a change in the toluene layer to a wine-red colour due to the formation of colloidal gold. The organic layer is then separated, washed free of inorganic salts and dried. Such solutions of what we refer to as "underivatised" particles are remarkably stable, showing no sign of agglomeration or precipitation of the metal over periods of many months. The solutions show a well defined absorption band at 530 nm characteristic of the surface plasmon resonance of bulk gold and indicating that the gold particles have diameters in the range 5–10 nm (Caro and Ingrand, 1965). Removal of the solvent by evaporation leads to precipitation of the metal particles and the quaternary ammonium salt. If the salt is removed, the residual metal nanoparticles quickly sinter and will not reform the colloidal solution on treatment with fresh toluene.

When, however, the tetrachloroaurate reduction is carried out with addition of an alkanethiol [molar ratio thiol: Au = *ca.* 1.3] immediately before the borohydride solution, the toluene layer becomes a dark brown colour. Washing to remove inorganic salts, drying and evaporation of part of the toluene followed by addition of ethanol affords a precipitate from which excess thiol can be washed with further ethanol. After drying, the residue can be taken up again in toluene, reforming the dark solution, which shows only a broad featureless absorption decreasing in the range 300–800 nm with no sign of the surface plasmon absorption at 530 nm. Such an appearance is indicative of much smaller metal particles than in solutions of the "underivatised" particles. Subsequent transmission electron microscopic examination showed that the thiol-capped particles were indeed of remarkably uniform size, with diameters in the range 1.5 to 3 nm. Further characterisation of the nonanethiol-derivatised particles confirmed that they consisted of gold, carbon hydrogen and sulfur only. Infrared spectra that matched the spectrum of the thiol except for the absence of the very weak S-H stretching band at around 2560 cm^{-1}. A ^{1}H-nmr spectrum had broadened signals in the range δ 1–2 corresponding to those in the spectrum of non-anethiol due to the protons attached at C-2 to C-9, but no signal at *ca.* δ 2.25 for the S-CH$_2$ group presumably because immobilisation rendered it too broad to be detected. (For a more detailed study of such effects, see Badia *et al.*, 1996a,b, 1997.) Mass spectrometric examination of the sample using a MALDI TOF spectrometer (by courtesy of Fisons Ltd.) showed that the sample consisted mainly of particles having four quite well defined m/z values 14 500, 21 600 and 38 400 with the dominant peak in the spectrum corresponding to m/z 29 500. This last value is close to that expected for Au$_{147}$ ($m/z = 28\,945$ without thiol; 35 019 assuming the presence of 30 thiolate moieties); here 147 is one of the so-called "magic numbers" referred to by cluster chemists in describing metal cluster compounds and would correspond to a cluster having 92 surface gold atoms believed (Sellers *et al.*, 1993) to be capable of binding a maximum of 30 thiolate groups in threefold hollow sites. (For an alternative discussion of particle size, see Whetten *et al.*, 1996)

SELF-ORGANISATION OF GOLD NANOPARTICLES

Electron microscopy of gold nanoparticles, both thiol-capped and "underivatised" particles has revealed that, when deposited from solution in toluene on to a glassy carbon coated copper grid for examination, they show a remarkable ability to organise themselves into ordered arrays. Figure 2 shows a micrograph of a single layer of "underivatised" particles prepared using tetraoctylammonium bromide self-organised into a pseudohexagonal raft.

When a second layer of particles is deposited on such hexagonal arrays, patterned areas appear which show characteristic projections of lines in three domains oriented in directions at angles of $120°$ to one another (Figure 3). Also in Figure 3 can be seen evidence for the deposition of some cyclic arrays, an effect that is very much enhanced in gold particles prepared in the presence of thiols, as typified by Figure 4 in which extended cyclic arrays can be seen. Such arrangements can be reproduced only if deposition of particles in the second layer takes place in the two-fold saddle sites between particles in the first layer rather than the threefold hollow sites that would maximise the number of interactions between particles in the two layers. This is illustrated in Figure 5(a) in which it can be seen that, for any particle in the hexagonal first layer, there exist three possible two-fold sites, designated A, B and C. The linear projections of Figure 3 arise from successive placements of second layer particles in AAA... or BBB... or

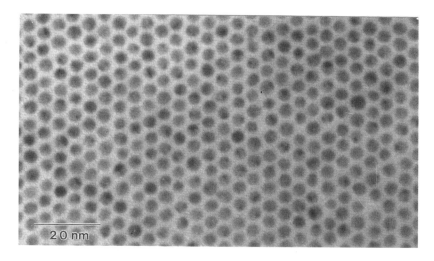

Figure 2 Electron micrograph showing a pseudohexagonal raft composed of a single layer of "underivatised" gold nanoparticles deposited by evaporation from toluene solution.

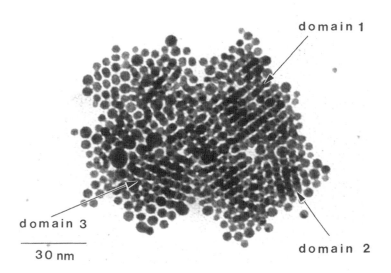

Figure 3 Electron micrograph of "underivatised" gold nanoparticles showing the effect of deposition of a second layer. The linear projections appear in three domains with directions at angles of 120°. A few cyclic domains can also be seen.

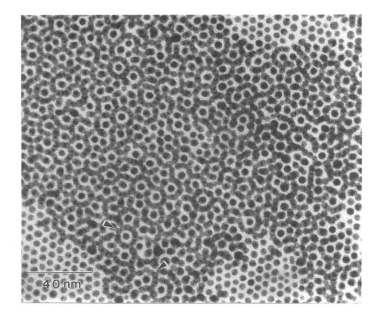

Figure 4 Electron micrograph showing cyclic domains formed by deposition from toluene solution of thiol-capped gold nanoparticles.

CCC...., while cyclic projections arise from sequences of placements of the type ABCABC as represented in Figure 5(b).

This preference for twofold sites can be understood in terms of a model for the "underivatised" particles as shown in Figure 6 in which the gold nanoparticle has adsorbed bromide ions on its surface, each associated with a counter cation, the tetra-alkylammonium ion; a somewhat similar interpretation was introduced by Bönnemann *et al.* (1991) and has been used by him and Reetz (Reetz and Helbig, 1994; Reetz and Quaiser, 1995) to explain the stability and solubility characteristics of electrochemically produced metal nanoparticles. On our interpretation, the surface of the gold nanoparticle is covered with an arrangement of ion pairs the packing density of which will be determined by the length of the alkyl chains of the ammonium ions. The electrostatic repulsion between contiguous particles in superposed layers will be minimised by occupancy of the twofold sites, but it will be offset to some degree by van der Waals attraction between the alkyl chains of the cations.

Evidence in support of this interpretation (which also explains the remarkable stability of toluene solutions of "underivatised" gold nanoparticles) comes from the following observations.

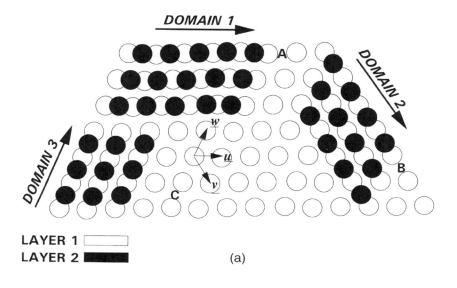

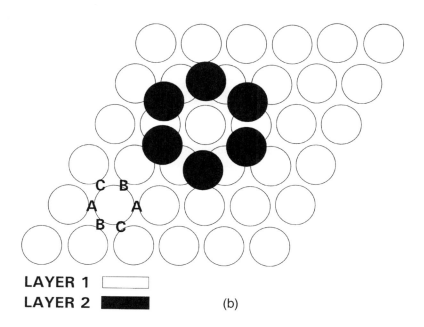

Figure 5 Schematic representation of the formation of the observed projections by deposition of the second layer of gold nanoparticles in the twofold saddle sites of the first layer: (a) linear domains; (b) cyclic domains.

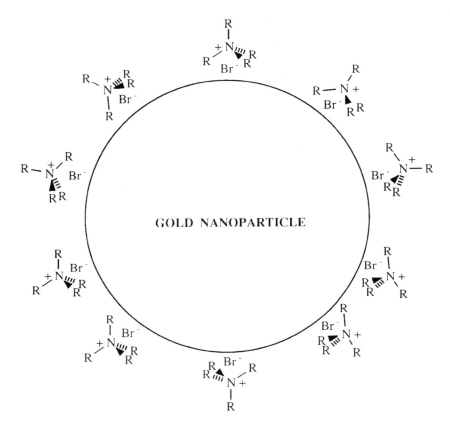

Figure 6 Proposed model of "underivatised gold nanoparticles. The surface area occupied by the attached quaternary ammonium bromide is governed in part by the spread of the R_3N^+- group, and the particle separation by the length of the alkyl chain R.

1. Energy dispersive X-ray (EDX) spectra determined on single layers of the "underivatised" particles show signals characteristic of Au (2.5 and 9.9 keV) and Br (1.9, 12.0, 13.0 and 13.5 keV). Examination of areas away from the particles showed no sign of Br.

2. The average separation in pseudohexagonal rafts of gold particles prepared using quaternary ammonium salts of varying chain length increases approximately linearly as the chain length increased from C6 to C18 (Figure 7). The magnitude of the range of separations suggests that the alkyl chains are not fully extended and/or are interdigitated with chains on ammonium ions associated with adjacent particles.

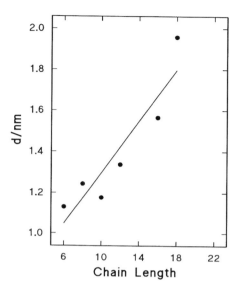

Figure 7 Dependence of the average separation (d/nm) of "underivatised" gold nanoparticles in arrays formed by evaporation from toluene solution on the alkyl chain length in the tetra-alkylammonium bromide used in their preparation.

These observations provide convincing evidence that "underivatised" gold nanoparticles are in fact associated with bromide ions and tetra-alkylammonium ions as required by the electrostatic interpretation of the organisation seen by electron microscopy.

Somewhat similar, though rather less well defined, examples of self organisation have been detected in alkanethiol-capped gold nanoparticles and in material obtained by α, ω-alkanedithiol treatment of solutions of "underivatised" particles (Brust *et al.*, 1995a). This may be interpreted again in terms of a balance between attractive dispersion forces that would favour occupancy of threefold hollow sites and electrostatic repulsion arising from the organised array of thiolate dipoles on the particle surfaces. In some cases, spectacular examples of size selection and ordering have been detected as illustrated in Figure 8 which shows the effect of adding excess decanethiol to a toluene solution of "underivatised" particles prepared using tetra-decylammonium bromide. Areas appear in which particles having diameters of some 8 nm are arranged in an hexagonal array, but with each particle surrounded by an hexagonal grouping of smaller particles having a diameter about half that of the larger ones. The origin of this effect, which seems to be rather specific for the combination of quaternary ammonium cation and thiol chain length, is unclear. The ability of adsorbates to restructure metal

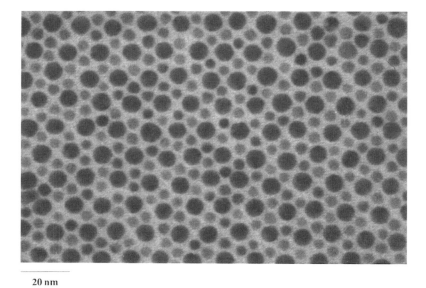

20 nm

Figure 8 Electron micrograph showing a bimodal hexagonal array formed by evaporation of a toluene solution of "underivatised" gold nanoparticles (prepared using tetradecylammonium bromide) to which had been added excess decanethiol.

surfaces is well known, and this might lead to changes in the size distribution of the original "underivatised" particles on attachment of thiolate ligands. At this time we have no convincing evidence in support of such an interpretation.

CONSTRUCTION OF TWO- AND THREE-DIMENSIONAL ARRAYS

Borohydride reduction of tetrachloroaurate by the standard two-phase procedure described earlier but using, in place of the alkanethiol, an α, ω-alkanedithiol led to formation of a precipitate. After washing and drying, electron microscopy of the amorphous dark material showed that it consisted of gold nanoparticles of 1.5–3 nm diameter held apart in an organic matrix presumably generated by cross-linking between adjacent gold particles. Analogous materials containing larger metal particles could also be made by addition of a toluene solution of a dithiol to a solution of "underivatised" gold nanoparticles; electron microscopy of samples of the mixture during the course of the precipitation of the cross-linked material indicated the generation of large aggregates showing strings and rings of particles rather similar in appearance to those obtained by evaporation of toluene solutions of "underivatised" particles and described above (Brust *et al.*, 1995a). Such methods of generation of arrays of gold nanoparticles was felt to be too uncontrolled for our purposes, although we did carry out conductivity measurements on compressed pellets of the material (see below). Accordingly we sought a more controllable procedure for organising arrays.

Again we chose a very simple method of layer-by-layer construction on a flat substrate such as a glass microscope slide or evaporated gold surface (Bethell *et al.*, 1996). The substrate was first treated so as to attach organic molecules bearing terminal thiol groups; clean glass surfaces were treated with trimethoxymercaptopropylsilane, $(MeO)_3SiCH_2CH_2CH_2SH$, while gold surfaces were simply dipped into toluene solutions of an α, ω-alkanedithiol. The resulting surface now fully covered with closely packed thiol groups was then dipped into a solution of "underivatised" gold nanoparticles in toluene for several hours. Thereafter the substrate, now bearing a coating of gold particles attached to the outer thiol groups, was washed free of unbound gold and then dipped into a solution of the dithiol so as to reform an outer organic layer again terminating in thiol groups. By alternate dipping in the colloidal gold solution and dithiol solution, layered materials could be built up in thicknesses that were limited only by the researcher's stamina and patience. The progress of layer deposition could be easily monitored on glass substrates by measurement of the absorbance in the visible region of the spectrum (Figure 9); the absorbance showed a linear

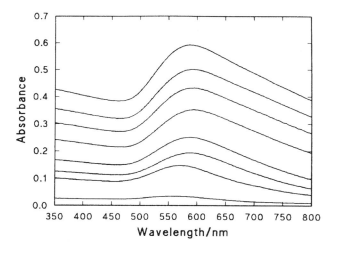

Figure 9 Visible absorption spectra of a glass substrate on the successive attachment of eight layers of "underivatised" gold nanoparticles using 1,9-nonanedithiol as the spacer unit.

increase with the number of layers, and this parallelled the total thickness of the applied material as measured ellipsometrically. With sufficient layers, the materials were dark in colour but displayed a faint metallic lustre. Indeed the strong absorbance at wavelengths above 700 nm in Figure 9 (not present in solutions of the metal particles) seems to suggest quasi-metallic electronic interaction between the metal particles in these layered arrays (compare Vossmeyer *et al.*, 1994; Musick *et al.*, 1997). Of course, such a simple method of layer construction leads to imperfections and these tend to be cumulative; an STM examination of the surface of a glass slide bearing 39 layers of gold particles confirmed this. The appreciable irregularity that was observed is thought to arise in part from imperfections in the surface of the substrate, incomplete derivatisation of the surface before the application of the first layer of particles and appreciable looping of the dithiol molecules so that both ends are attached to the same gold particle. The main source of imperfections, however, is probably the size variation among the metal particles. Further advances in this technique will require the development of methods of separation of preparations of nanoparticles into fractions having even narrower size distributions; Whetten and his co-workers (Whetten *et al.*, 1996) have already demonstrated that such fractionation is possible and can lead to crystalline superlattices.

PROPERTIES OF THE MATERIALS

With the capability of making macroscopic arrays of nanoparticles, albeit structurally rather imperfect, we have undertaken investigation of some of their electronic and optoelectronic properties. Thus electrical conductivity, photoconductivity, electrolyte electroreflectance and electron transfer properties have been studied. The picture that emerges is of a family of materials the properties of which can be controlled and indeed tailored by the choice of metal particle size and by the nature of the organic molecules that act as spacers between them.

Electrical Conductivity

Electrical resistance measurements were carried out on materials prepared using the layer-by-layer method and therefore containing gold particles with diameters in the range 5–10 nm. Specific resistivities for materials prepared with α, ω-hexane-, nonane- and dodecane-dithiols increased by roughly an order of magnitude for each increase in the alkane chain by three methylene groups.

A greater diversity of metal particle size and dithiol structure could be examined by using compressed pellets made from precipitated material (Brust *et al.*, 1995a). The material was loaded into precision bore glass capillaries between two tightly fitting silver steel wire electrodes, between which the sample was compressed mechanically to 2 bar, and the d.c. resistance measured, together with the length of the compressed sample (travelling microscope). For all the samples, specific conductivities (σ), determined over a range of temperatures, increased with increasing temperature, indicating nonmetallic conductivity and could be fitted to equation (1) and an apparent activation energy determined graphically. This behaviour is consistent with an activated electron hopping mechanism for which Abeles *et al.* (1975) have derived the expression (2) on the basis of simple electrostatic considerations.

$$\sigma = \sigma_0 \exp(-E_a/RT) \qquad (1)$$

$$E_a = \frac{1}{2} \frac{e^2}{4\pi\varepsilon\varepsilon_0} \left[\frac{1}{r} - \frac{1}{r+s} \right] \qquad (2)$$

In (2), r represents the radius of the metal particle and s is the length of the organic spacer between the particles, estimated for present purposes from molecular modelling of the most extended conformation of the polymethylene chain plus an estimate of the C–S–Au distance taken from the work of Sellers *et al.* (1993). The results in Table 1 demonstrate that conductivity increases with increasing gold particle size and decreasing distance between the thiol functionalities in the spacer. The observed activation barriers to electron hopping are reproduced quite well by (2), supporting the interpretation. The possibility now exists to design conductors with predetermined electrical characteristics.

Table 1 Specific conductivities and activation barriers in dithiol-linked gold nanoparticles.

Dithiol	$2r$/nm[a]	s/nm[b]	$\sigma/\Omega^{-1}\,cm^{-1}$	E_a/eV Observed	E_a/eV Equation (2)
HS-(CH$_2$)$_5$-SH	2.2	1.114	6.7×10^{-6}	0.14	0.165
HS-(CH$_2$)$_6$-SH	2.2	1.295	3.6×10^{-7}	0.12	0.177
HS-CH$_2$-C$_6$ H$_4$-CH$_2$-SH	2.2	1.075	6.7×10^{-6}	0.10	0.162
	8	1.075	7.7×10^{-2}	0.020	0.019
HS-(CH$_2$)$_{12}$-SH	8	2.047	1.3×10^{-3}	0.030	0.030

[a] Mean diameter of gold particles.
[b] Maximum distance between gold particles computed from the dithiol structure in its most extended conformation.

Photoconductivity

Experiments have been carried out in Liverpool in which gold nanopaticles have been laid down layer-by-layer on a p-doped silicon substrate using α, ω-nonanedithiol as the spacer and covered by an evaporated layer of metallic gold. Irradiation of the layers through the upper gold film while scanning the applied potential difference across the layers led to the observation of photocurrents only at potentials below $-0.4\,V$ as found in the absence of the nanostructured material. A similar experiment on a similar 6-layer assembly in which the simple alkanedithiol had been replaced by $HS\text{-}(CH_2)_8\text{-}Fc\text{-}(CH_2)_8\text{-}SH$ (Compound **1**), a dithiol containing a fer-rocene (Fc) moiety in the middle of the methylene chain, also gave rise to photocurrents at potentials less than $-0.4\,V$ but, significantly, also pro-duced substantial photocurrents at potentials greater than $+0.15\,V$. Much further work is necessary, but this simple result again emphasises how the properties of these nanostructured materials can be changed by the choice of the structure of the organic (or in this case organometallic) spacer.

Electrolyte Electroreflectance (Baum and Dougherty, 1996)

Four layers of gold nanoparticles were assembled on a polycrystalline gold electrode using 1,9-nonanedithiol as the spacer between the layers. At electrode potentials in the range -0.1 to $-0.3\,V$ (vs. SCE), strong electro-chemical modulation of the reflectivity was observed at about $500\,nm$ near the gold surface plasmon band, the signals being shifted both in phase and wavelength compared with a clean gold surface. Changing the spacer dithiol to **1** led to a pronounced, potential-dependent shift in the reflectance response to longer wavelengths (*ca.* $560\,nm$ at $-0.3\,V$ and 580 at $+0.4\,V$. The mechanism of these changes in response are not yet understood, but the observations seem to hold out the promise of developing ultra-thin films of these nanostructured materials for the controlled modification of the reflectivity of surfaces.

Heterogeneous Electron Transfer (Brust, 1995)

The reversible potassium ferri-/ferro-cyanide couple in aqueous solution was used as the probe electron-transfer process. Starting from a clean gold spherical electrode of $0.2\,cm^2$ area, the cyclic voltammetric (CV) response was measured as the electrode surface was modified, first with a layer of 1,9-nonanedithiol, then with a layer of "underivatised" gold nanoparticles, then with a second layer of dithiol, then with a further layer of gold nanopar-ticles and so on. In each case the presence of the dithiol as the outer layer completely suppressed the CV response, which was restored by the attachment of an outer layer of gold particles. We found this behaviour remarkable in that, transmission of electrons can clearly take place through the alternating layers of dithiol and gold paticles, but heterogeneous electron transfer from the surface is apparently only possible from surface gold particles. In order to separate the effect on the rate of heterogeneous electron transfer (characterised by the rate constant k_{het}) as the number of layers increased from the concomitant increase in resistance, AC impedance spectroscopy (5000 to 1 Hz) was used. The derived values of k_{het} showed dramatic fluctuations depending on whether the electrode's outer surface is composed essentially of gold or thiols as indicated schematically in Figure 10. It may be noted that the value of k_{het} for the system with gold particles as the outer layer showed a progressive decline with increasing numbers of layers of gold particles. We cannot provide a complete explanation of our observations, but we feel that they, like the visible spectra of the layers, may indicate that the array of nanoparticles may possess electronic interactions that give a quasi-metallic band structure to the electronic levels of the electrode coating.

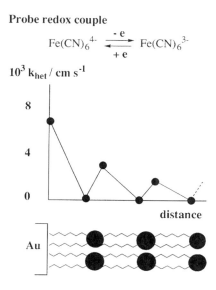

Figure 10 Schematic representation of the rate constants for heterogeneous electron transfer (k_{het}) from a gold electrode bearing on its surface alternate layers of 1,9-nonanedithiol and "underivatised" gold nanoparticles. Points correspond to values measured as the surface layer alternates between gold nanoparticles and nonanedithiol.

FUTURE DEVELOPMENTS

Our investigations in the area of gold nanoparticles and their arrays within organic matrices convince us that materials derived from them will prove a valuable stepping stone towards the goal of truly molecular electronics and also find diverse applications in other fields from physics to medicine. Further progress in the development of the basic science of these materials can be anticipated in a number of areas:

1. Extending the range of nanoparticles. Already nanoparticles of other (transition) metals are being actively investigated (see, for example, Reetz et al., 1997) as are the possibilities of using semiconductors such as silicon (see, for example, Burr et al., 1997; Koch and Petrova-Koch, 1996) and cadmium sulfide (Colvin et al., 1992; Brust et al., 1996; Braun et al., 1996).

2. Incorporation of more diverse organic structures into the spacer units could render the materials (as thin films) sensitive in a spatially resolved way to a wide range of external inputs such as one electron transfer, electromagnetic radiation, cations and anions.

3. Important advances can be expected from the development of reliable methods of producing monodisperse particles. Particle produced by existing methods, although having quite narrow size distributions, cannot exploit fully the sensitive dependence of properties on size. The important work of Whetten, who used a fractional precipitation method to separate preparations of thiol-capped gold particles, enabled him to crystallise "nanocrystal superlattices", as confirmed by X-ray diffraction. (Whetten et al., 1996)

4. Improved methods of ordering arrays of nanoparticles can be anticipated. The simple but as yet rather poorly controlled methods of assembly that we have developed can clearly be improved, and we have previously suggested protocols by which chemical reactions of functional groups within the organic spacers might be used to define the structure of an array (Brust et al., 1995b; Bethell et al., 1996). Already, however, sophisticated methods of assembly using interactions from the realm of biology have been adopted. For example, polynucleotide base-pair interactions have been applied (Alivisatos et al., 1996; Mucic et al., 1996) using gold nanoparticles tagged with a polynucleotide sequence that can be recognised by a DNA template, hence locating the gold particles. Further exciting developments will surely follow.

REFERENCES

Abeles, B., Cheng, P., Coutts, M.D. and Arie, Y. (1975). *Adv. Phys.*, **24**, 407.

Alivisatos, A.P., Johnsson, K.P, Peng, X., Wilson, T.E., Loweth, C.J., Bruchez Jr. M.P. and Schulz, P.G. (1996). *Nature*, **382**, 609.

Badia, A., Gao, W., Singh, S., Demers, L., Cuccia, L. and Reven, L. (1996a). *Langmuir*, **12**, 1262.

Badia, A., Singh, S., Demers, L., Cuccia, L., Brown, G.R. and Lennox, R.B. (1996b). *Chemistry — A European Journal*, **2**, 359.

Badia, A., Cuccia, L., Demers, L., Morin, F. and Lennox, R.B. (1997). *J.Am. Chem. Soc.*, **119**, 2682.

Baum, T. and Dougherty, G. (1996). Unpublished experiments.

Bönnemann, H., Brijoux, W., Brinkmann, R., Dinjus, E., Joussen, T. and Korall, B. (1991). *Angew. Chem. Int. Ed. Engl.*, **30**, 1312.

Braun, P.V., Osenar, P. and Stupp, S.I. (1996). *Nature*, **380**, 325.

Brust, M. (1995). Ph.D. Thesis, University of Liverpool.

Brust, M., Walker, M., Bethell, D., Schiffrin, D.J. and Whyman, R. (1994). *J. Chem. Soc. Chem. Comm.*, 801.

Brust, M., Bethell, D., Schiffrin, D.J. and Kiely, C.J. (1995a). *Adv. Mat.*, **7**, 795.

Brust, M., Fink, J., Bethell, D., Schiffrin, D.J. and Kiely, C.J. (1995b). *J. Chem. Soc. Chem. Comm.*, 1655.

Brust, M., Etchenique, R., Calvo, E.J. and Gordillo, G.J. (1996). *J. Chem. Soc. Chem. Comm.*, 1949.

Burr, T.A., Seraphin, A.A., Werwa, E. and Kolenbrander, K.D. (1997). *Physical Review B — Condensed Matter*, **56**, 4818.

Colvin, V.L., Goldstein, A.N. and Alivisatos, A.P. (1992). *J. Am. Chem. Soc.*, **114**, 5221.

Hostetler, M.J., Green, S.J., Stokes, J.J. and Murray, R.W. (1996), *J. Am. Chem. Soc.*, **118**, 4212.

Koch, F. and Petrova Koch, V. (1996). *Journal Of Non-Crystalline Solids*, **200**, 840.

Mucic, R.C., Stornhoff, J.J., Letsinger, R.L. and Mirkin, C.A. (1996). *Nature*, **382**, 607.

Musick, M.D., Keating, C.D., Keefe, M.H. and Natan, M.J. (1997). *Chem. Mater.*, **7**, 1499.

Reetz, M.T. and Helbig, W. (1994). *J. Am. Chem. Soc.*, **116**, 7401.

Reetz, M.T. and Quaiser, S.A. (1995). *Angew. Chem. Int. Ed. Engl.*, **34**, 2240.

Reetz, M.T., Winter, M. and Tesche, B. (1997). *J. Chem. Soc. Chem. Comm.* 147.

Sellers, H., Ulman, A., Shadman, Y. and Eilers, J.E. (1993). *J. Am. Chem. Soc.*, **115**, 9389.

Terrill, R.H., Postlethwaite, T.A, Chen, C.-H., Poon, C.-D., Terzis, A., Chen, A., Hutchinson, J.E., Clark, M.R., Wignall, G., Londono, J.D., Superfine, R., Falvo, M., Johnson Jr. C.S., Samulski, E.T. and Murray, R.W. (1995). *J. Am. Chem. Soc.*, **117**, 12537.

Whetten, R.L., Khoury, J.T., Alvarez, M.M., Murthy, S., Vezmar, I., Wang, Z.L., Stephens, P.W., Cleveland, C.L., Luedtke, W.D. and Landman, U. (1996). *Adv. Mater.*, **8**, 428.

13. ULTRAFAST OPTICAL CONTROL OF ION PAIR INTERMEDIATES IN ELECTRON TRANSFER REACTIONS: TOWARD A MOLECULAR SWITCH**

MICHAEL R. WASIELEWSKI[a,b,]*, MARTIN P. DEBRECZENY[b],
EMILY M. MARSH[a] and WALTER A. SVEC[b]

[a]*Department of Chemistry, Northwestern University, Evanston,
Il 60208-3113, USA*
[b]*Chemistry Division, Argonne National Laboratory, Argonne,
IL 60439-4831, USA*

ABSTRACT

Three approaches to the design, synthesis, and characterization of molecular switches are described. These strategies make use of high quantum yield, picosecond time scale electron transfer processes to control the movement of electrons within molecules with the ultimate goal of developing new opto-electronic materials for data manipulation. These approaches include the use of photogenerated electric fields to control electron transfer rates, direct photonic control of ion pair intermediates, and bridging group photo-activation.

INTRODUCTION

The design and production of state-of-the-art electronic and opto-electronic devices depends increasingly on the ability to produce ever-higher densities of circuit elements within integrated circuits. (Miller, 1990a) As technical progress reduces component sizes closer to molecular dimensions, it is clear that new and fundamental research into the distinctly different physical phenomena that may occur in molecular electronics as compared to larger

---

* To whom correspondence should be addressed.
** The submitted manuscript has been created by the University of Chicago as Operator of Argonne National Laboratory ("Argonne") under Contract No. W-31-109-ENG-38 with the U.S. Department of Energy. The U.S. Government retains for itself, and others acting on its behalf, a paid-up, nonexclusive, irrevocable worldwide license in said article to reproduce, prepare derivative works, distribute copies to the public, and perform publicly and display publicly, by or on behalf of the Government.

scale electronics is necessary. The successful development of new molecule-sized electronics promises to yield dramatic improvements in component density, response speed, and energy efficiency. In the component density area alone, molecular switches may increase data storage density to as much as 10^{18} bits cm^{-2}. (Carter, 1982) By comparison, current microfabrication techniques have made $0.5\,\mu$m component sizes routine in semiconductor materials, which produce a data storage density of 1.6×10^7 bits cm^{-2}. Massive size reductions using molecular switches will most likely be limited by quantum statistical considerations, if reasonable data error rates are to be maintained. (Haddon and Lamola, 1985; Miller, 1990b; Miller, 1990c) Nevertheless, molecular devices that use visible light for addressing and control purposes have a realizable data density of $\sim 2 \times 10^9$ bits cm^{-2} just through diffraction limited spot size considerations alone. Recent results suggest that three dimensional addressing, (Parthenopoulos and Rentzepis, 1989) the use of excitonic waveguides, (Lieberman *et al.*, 1990) and near-field (Feldstein *et al.*, 1996; Higgins *et al.*, 1996) optical techniques can greatly increase this resolution. Furthermore, since energy and electron transfer processes within molecules can take place on a sub-picosecond time scale, it is possible to produce devices that respond far more rapidly than current devices. Another advantage of using light-driven molecular devices is the ability to perform optical multiplexing, whereby light of specific frequencies is used to selectively address individual components. (Miller, 1990c) Finally, employing high quantum efficiency, fast photo-driven processes will decrease the heat load produced by molecular devices. This results in a much more energy-efficient and reliable system. These latter advantages buttress the higher data densities cited above, and may result in economically viable molecular electronic devices.

Recent efforts in the field of molecular electronics have been most heavily concentrated on molecular wires, chemical sensors, which include single-molecule detection, and non-linear optical polymers for switching and modulation. Molecular wire research has focused mainly on conjugated oligomers and polymers (Kemp *et al.*, 1996; Patil, Heeger and Wudl, 1988; Reimers *et al.*, 1996; Schumm, Pearson and Tour, 1994; Shimidzu *et al.*, 1995; Tolbert, 1992; Wagner and Lindsey, 1994). Low dimensional materials derived from these conjugated molecules can be doped to make them highly conductive. Films of these materials are finding applications as diverse as organic batteries (Alcacer, 1987) and electrochromic devices. (Burroughes *et al.*, 1990) Since the conductivity of these materials is very sensitive to chemical doping, electron donor or acceptor molecules have been attached to these polymers to modulate their dopant level, and thereby change their conductivity. (Baeuerle *et al.*, 1990) In addition, polyarene ladder polymers (Yu, Chen and Dalton, 1990) and conductive polyimides (Dietz *et al.*, 1990) are useful both as conductive materials and as non-linear optical (NLO) materials.

Chemical sensors have been developed that interface crown ethers and other molecules with ion or molecular recognition properties to electronic devices such as field effect transistors. (Reinhoudt and Sudhoelter, 1990) The ability of one molecule to selectively recognize another is providing the basis for developing practical molecular sensors. (Fouquey, Lehn and Levelut, 1990; Swager and Marsella, 1994; Willner and Willner, 1997) Molecular motion and recognition also play a combined role in several such systems. (Asakawa *et al.*, 1996; Marsella, Carroll and Swager, 1994) Optical multiplexing has been demonstrated using two-color photon-gated hole burning in electron donor-acceptor molecules. (Carter *et al.*, 1987) In this work narrow frequency lasers are used to write and address information in a matrix of donor-acceptor molecules at low temperature. In addition single molecule spectroscopy has been used to develop molecular switches. (Moerner *et al.*, 1994)

Molecule-based optical switching research has focused intensively on the preparation of polymers with NLO properties. (Blanchard-Desce, Marder and Barzoukas, 1996; Prasad and Reinhardt, 1990) Non-linear optical response can be imparted to a polymer either by its intrinsic chemical structure or by appending organic molecules with known NLO properties to preformed polymer backbones. These polymers are finding application in frequency conversion of light through second and third order non-linear processes. Using these materials in practical devices depends to a large extent on finding ways to deal with materials problems such as solubility, optical transparency, and mechanical properties.

The relatively new field of molecular electronics has produced promising results for switches and opto-electronic components. (Birge, 1995; Carter, 1982; Carter, 1987; Carter, Siatkowski and Wohltjen, 1988; de Silva and McCoy, 1994; DelMedico *et al.*, 1995; Feringa, Huck and Schoevaars, 1996; Gomez-Lopez, Preece and Stoddart, 1996; Lehn, 1988; Mehring, 1989; Mirkin and Ratner, 1992; Tour, 1996; Willner and Willner, 1997). The earliest ideas regarding the possibility of producing molecule-sized optical switches were based on using an electron transfer reaction within a donor-acceptor pair. (Aviram and Ratner, 1974) Subsequent theoretical discussions strongly support the idea that an electron transfer reaction can form the basis for a molecular switch. (Aviram and Ratner, 1974; Hopfield, Onuchic and Beratan, 1988; Metzger, 1994; Waldeck and Beratan, 1993) Yet, the design, synthesis, and demonstration of such switches has been slow to develop because the optimal use of electron transfer processes requires careful control of molecular structure, electronic coupling, and thermodynamics in a complex array of donors and acceptors. Experimental work has focused mainly on approaches to molecular switches that make use of photochemical and electrochemical conformational transformations within molecules. (de Silva and McCoy, 1994; Feringa *et al.*, 1996; Gomez-Lopez *et al.*, 1996; Mirkin and Ratner, 1992; Tour, 1996; Willner and Willner, 1997) Our work in the field of photoinduced electron transfer reactions affords us a unique opportunity to exploit the information and techniques, which we and others have developed, to investigate new approaches to molecular electronic devices. (Gust, Moore and Moore, 1993; Kurreck and Huber, 1995; Wasielewski, 1992).

RESULTS AND DISCUSSION

Our fundamental approach to achieve efficient, long-lived photoinitiated charge separation is to produce arrays of electron donor-acceptor molecules in which the redox partners are rigidly attached in specific spatial arrangements relative to one another. (Gaines *et al.*, 1991; Wasielewski, 1992; Wasielewski *et al.*, 1988) Thus, photoexcitation of a donor or acceptor will result in vectorial electron transport to the redox partners attached to the photoexcited molecule, provided that the free energies and electronic coupling matrix elements of the electron transfer reactions taking place within them are optimized. Electron transfer theory provides a framework for understanding how to carry out this optimization. (Jortner, 1976; Marcus, 1965) These are fundamental requirements for any electron donor-acceptor molecule that forms the basis of a molecular electronic device. Until recently, what has been lacking is the ability to control the rates of formation and decay of the various ion pair intermediates within an extended donor-acceptor array. We have developed three approaches to controlling electron transfer rates to yield specific ion pair intermediates so that they can function as molecular logic elements. Work in progress to establish these approaches is detailed below.

One method of controlling the rates of electron transfer reactions is through the application of electric fields. As a consequence, a considerable amount of work has been devoted to the theoretical modeling (Aviram, Joachim and Pomerantz, 1988; Aviram and Ratner, 1974; Broo, 1993) and experimental realization (Garnier *et al.*, 1990; Sambles and Martin, 1993; Tour, Wu and Schumm, 1991) of molecular electronic switches consisting of organic electron donor-acceptor pairs whose operation is controlled by an external electric field. Langmuir-Blodgett films containing monolayers of donor and acceptor chromophores have been created in which control of electron transfer was achieved by varying the layer composition separating the donors and acceptors and by the application of external fields. (Kuhn, 1985; Wilson, 1995) Recently, it was shown that the electric dipole of a synthetic α-helical polypeptide can influence the rate of electron transfer between organic electron donors and acceptors covalently attached to the polypeptide. (Galoppini and Fox, 1996) Thus, the electric field produced by a photogenerated ion pair can have a large effect on the electronic states of surrounding molecules. We have studied donor-acceptor-probe molecular arrays in which photogeneration of donor$^+$-acceptor$^-$ ion pairs occur in less than 10 ps in low polarity solvents. (Debreczeny, Svec and Wasielewski, 1996b; Gosztola, Yamada and Wasielewski, 1995) These ion pairs produce an $\sim 6\,MV/cm$ electric field that shifts the optical absorption spectra of polarizable probe molecules by 5–15 nm. These results suggest that it might be possible to use the large, anisotropic, local electric fields generated by the formation of ion pairs to control a second photoinduced electron transfer reaction on a picosecond time scale. Generation of an electric field at the molecular level using photogenerated ion pairs holds several advantages over external, macroscopic field generation: (1) the applied field strength can be larger because macroscopic field strengths are limited by dielectric breakdown, (2) the field can be turned on and off on a picosecond time scale, (3) the electric field is under optical control, and (4) only a very small volume is affected by the locally applied electric field.

We have prepared a donor$_1$-acceptor$_1$-acceptor$_2$-donor$_2$ molecular array (D_1-A_1-A_2-D_2, **1**) (Debreczeny *et al.*, 1996b) in which photoinduced charge separation within one donor-acceptor pair controls the rate constants for photoinduced charge separation and thermal charge recombination within a second donor-acceptor pair. Multiple laser pulses with 130 fs duration are used to selectively control and probe the generation of the two ion pairs. The D_1-A_1-A_2-D_2 molecule needs to fulfill several major requirements to make it possible to observe the electric-field-induced switching effect. These requirements are general for this type of molecular switch. First, it must be possible to selectively excite the two donors, D_1 and D_2. Zinc 5-phenyl-10,15,20-tri(n-pentyl)porphyrin (D_1) and the phenyldimethyl-pyrromethene dye (D_2) were chosen for this purpose because they can be independently excited at 416 nm and 512 nm, respectively. Second, the two acceptors, A_1 and A_2, should not absorb light significantly at the two excitation wavelengths in both their neutral and singly reduced states. However, they should possess strong absorptions at other wavelengths that are independently observable when A_1 and A_2 are reduced. 1,4 : 5,8-Naphthalenediimide (A_1) and pyromellitimide (A_2) were chosen because they do not absorb significantly at 416 nm and 512 nm, and their radical anion spectra have well-separated, intense absorptions at 480 nm and 713 nm, respectively, Figure 2. (Osuka, 1993; Viehbeck, Goldberg and Kovac, 1990) The large extinction coefficients of these reduced acceptors provide distinct observables that are critical to their application within molecular switches. Third, the population of field generating ion pairs, D_1^+-A_1^-, within D_1-A_1-A_2-D_2 must be large enough to significantly affect the electron transfer reactions of A_2 - D_2. In **1** this was achieved by choosing a porphyrin that possesses a large absorption cross section at the 416 nm excitation wavelength ($\varepsilon_{416} = 10^5\,M^{-1}\,cm^{-1}$)

as D_1. (Gouterman, 1978) Fourth, redox potentials for the one-electron oxidation of D_1 and D_2, as well as the one-electron reduction of A_1 and A_2 must be chosen to ensure that charge separation and recombination occurs only within each individual donor-acceptor pair, D_1-A_1 and A_2-D_2. Fifth, the free energies for photoinduced charge separation from the lowest excited singlet states of the donors within each pair to their corresponding acceptors should be sufficiently negative to ensure rapid rates of ion pair formation in the low polarity solvents necessary to support large electric fields. The redox potentials for D_1, A_1, A_2, and D_2, as well as the lowest excited singlet state energies for D_1 and D_2 fulfill the fourth and fifth requirements. (Debreczeny *et al.*, 1996b)

Single pulse excitation of D_1-A_1-A_2-D_2 in 1,4-dioxane at either 416 nm ($^{1*}D_1$) or 512 nm ($^{1*}D_2$) led to ion pair formation of either D_1^+-A_1^- or A_2^--D_2^+, respectively, as was monitored at 480 nm (A_1^-) or 713 nm (A_2^-) (Figure 1). The time constants for ion pair formation and decay of either D_1^+-A_1^- or A_2^--D_2^+ within D_1-A_1-A_2-D_2 were similar to those observed for the D_1-A_1 and A_2-D_2 reference compounds. (Debreczeny *et al.*, 1996b) In the two pump pulse experiment, a single 512 nm pulse again produced D_1-A_1-A_2^--D_2^+. The transient absorption signal for A_2^- within D_1-A_1-A_2^--D_2^+ reached a maximum at about 700 ps after the arrival of the 512 nm pulse. At that time, a second 416 nm pulse was used to produce $^{1*}D_1$-A_1-A_2^--D_2^+, which underwent rapid electron transfer from $^{1*}D_1$ to yield D_1^+-A_1^--A_2^--D_2^+.

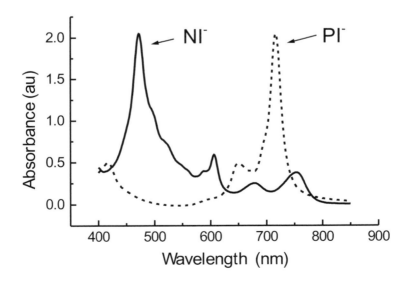

Figure 1   Ground state spectra of NI$^-$ and PI$^-$.

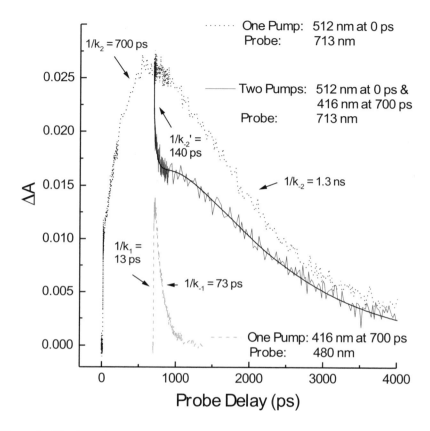

Figure 2 The Transient absorption kinetics of D_1-A_1-A_2-D_2 excited with one pump pulse at 416 nm, one pump pulse at 512 nm, and with two sequential pump pulses at 416 nm and 512 nm at the times indicated.

The population of A_2^- is selectively monitored at 713 nm (solid curve in Figure 2) so that the effect of the electric field generated by the D_1^+-A_1^- state on the population of the A_2^--D_2^+ state is observed. The data in Figure 2 show that the slow formation and decay of the A_2^--D_2^+ state is disrupted during the short lifetime of the D_1^+-A_1^- state because the electric field generated by the D_1^+-A_1^- state makes recombination of the A_2^--D_2^+ state to the ground state considerably faster ($\tau = 140$ ps) than its natural decay time ($\tau = 1.3$ ns). The rapid creation of the D_1^+-A_1^- dipole modifies the electronic environment in the vicinity of A_2^--D_2^+. These environmental changes can be imposed in a rapid and controlled fashion using ultrafast laser pulses. Our results suggest that this type of molecular architecture has many of the characteristics required of a prototype molecular switch, the most notable being the rapid control of state switching using femtosecond optical pulses.

Another method to implement rapid control of the on/off "states" of a molecular switch involves direct optical perturbation of one of the ions within a photogenerated ion pair intermediate. We have measured the two-pulse femtosecond electron transfer dynamics of a donor-acceptor (1)-acceptor (2) triad, **2**, in which an initial femtosecond laser flash is used to produce an ion pair, whose subsequent fate is controlled by a second femtosecond laser flash. (Debreczeny *et al.*, 1996a) The energetics of charge separation for **2** in toluene are given in Figure 3. (Greenfield *et al.*, 1996) Transient absorption measurements on **2** show that direct excitation of the charge transfer state of the ANI chromophore using 130 fs, 416 nm laser pulses results in electron transfer from ^{1*}ANI to the adjacent NI acceptor with $\tau = 420$ ps and a quantum yield of 0.95. The ANI$^+$-NI$^-$-PI state of **2**, which lives for $\tau = 20$ ns, exhibits characteristic optical absorptions for NI$^-$ at 480 nm and 605 nm, Figure 1. (Greenfield *et al.*, 1996; Viehbeck *et al.*, 1990; Wiederrecht *et al.*, 1996) The optical absorption spectrum of PI$^-$ has a strong band at 710 nm, while that of NI$^-$ at this wavelength is substantially weaker. (Osuka *et al.*, 1991; Viehbeck *et al.*, 1990; Wiederrecht *et al.*, 1996) There is no evidence for "uphill" electron transfer from NI$^-$ to PI within **2**, following single pulse excitation at 416 nm and probing at 710 nm.

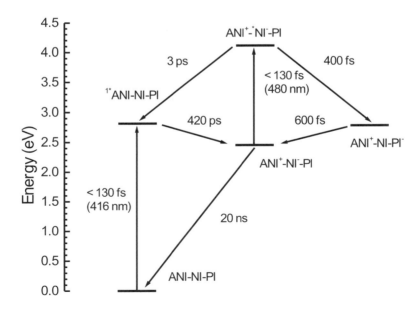

Figure 3   Energy level diagram for **2**.

However, a second pump pulse at 480 nm can provide the necessary energy to move the electron from NI^- to PI in compound **2**.

It is well known that photoexcitation of organic radical ions leads to the formation of excited states that may transfer an electron to nearby electron acceptors or to the solvent. (Eriksen, 1988; Scaiano *et al.*, 1988) The strong 480 nm band of NI^- allows us to selectively excite NI^- within ANI^+-NI^--PI to its lowest excited state, which in turn, permits us to observe competitive routes for deactivation of this excited radical anion state. By design, the electronic coupling between the states of NI and PI is greater than that between the states of ANI and NI because NI and PI lack the phenyl spacer between them. In **2**, the excitation of ANI at 416 nm results in the formation of ANI^+-NI^--PI. Three nanoseconds later, $^*NI^-$ is populated with the 480 nm pulse. Figure 4 shows the two-pump transient absorption spectrum of **2** measured 800 fs after the application of the 480 nm pump pulse. The 710 nm absorption of PI^- is clearly evident in the transient spectrum of **2**. The inset to Figure 4 shows the transient absorption kinetics at 710 nm for **2**. These data are compared with those for molecule **3**, which lacks the PI acceptor. Excitation of NI^- with a 480 nm pulse that occurs 3 ns following formation of ANI^+-NI^- in **3** with a 416 nm pulse results only in back electron transfer to produce $^{1*}ANI-NI$ with $\tau = 3$ ps. We have observed a 244 ps excited state lifetime for $^*NI^-$ in the absence of quenchers. In addition, NI^- shows no tendency to decompose following optical excitation with 480 nm , 130 fs laser pulses at a 2 kHz repetition rate.

The large positive absorption change for **2** at 710 nm indicates that electron transfer occurs from $^*NI^-$ to PI with $\tau = 400$ fs, while the return electron transfer from PI^- to NI occurs with $\tau = 600$ fs. The rate of electron transfer from $^*NI^-$ to PI is 8 times faster than the competitive electron transfer from $^*NI^-$ to ANI^+ and occurs with an quantum yield of 0.84. (Debreczeny *et al.*, 1996a) Thus, the second photon rapidly switches the charge separated state from ANI^+-NI^--PI to $ANI^+-NI-PI^-$ These experiments show that it is possible to control the movement of electrons within a covalently-linked array with multiple donor and acceptor sites on a femtosecond time scale.

In a donor-bridge-acceptor molecule the electronic coupling between the oxidized donor and reduced acceptor is dependent on the electronic structure of the bridging molecule. If the electronic structure of this molecule can be controlled, then the electronic coupling between donor$^+$ and acceptor$^-$ can be controlled. This control can be exerted by placing the bridging molecule into an excited state. (Mehring, 1989) In its lowest excited singlet state the HOMO and LUMO of the bridging molecule are singly occupied, and thus, the bridging molecule can act as an electron donor or acceptor, Figure 5. We have investigated this idea using molecule **4**, ZnP-PMI-NI. This molecule is designed so that the intermediate electron carrier, the

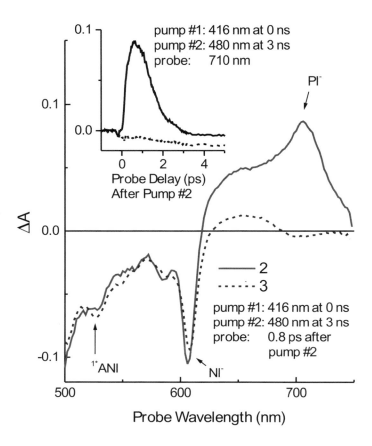

Figure 4 Two-pulse transient absorption spectrum of **2** and **3** in toluene. Inset: Transient absorption kinetics of **2** and **3** in toluene following two-pulse excitation.

3-phenylperylene-9,10-imide (PMI) is also capable of being independently excited following formation of the initial ion pair, ZnP^+-PMI-NI^-. Excitation of **4** with a 130 fs, 420 nm laser pulse results in formation of ZnP^+-PMI^--NI with $\tau = 2\,ps$, Figure 6. Subsequent thermal electron transfer from PMI^- to the terminal acceptor NI occurs with $\tau = 10\,ps$. The resultant ion pair exhibits a lifetime of $\tau = 2\,ns$ in THF solvent. Given the reasonably long lifetime of ZnP^+-PMI-NI^-, the perylene chromophore is returned to its ground electronic state and strongly absorbs at 502 nm. When PMI within ZnP^+-PMI-NI^- is excited with a 530 nm, 130 fs laser flash that occurs 30 ps following the initial 420 nm laser flash that creates

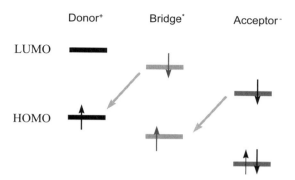

Figure 5   Energy level diagram.

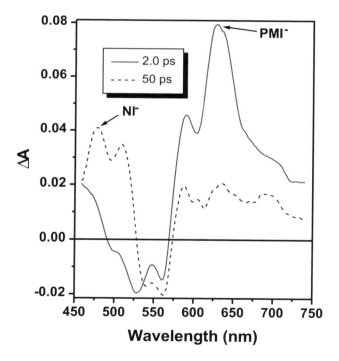

Figure 6   Transient absorption data for 4 in THF.

ZnP$^+$-PMI-NI$^-$, the transient behavior shown in Figure 7 is observed. At times < 1 ps following the 530 nm laser flash the 480 nm absorption due

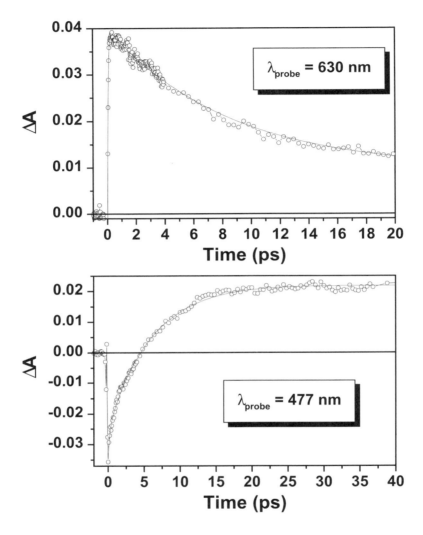

Figure 7 Transient absorption kinetics following two excitation pulses, 420 nm at $t = -30$ ps, and 530 nm at $t = 0$ ps.

to NI^- is strongly diminished and subsequently recovers with $\tau = 10$ ps. At the same time, the absorption of PMI^- reappears in < 1 ps, only to disappear with $\tau = 10$ ps. Thus, the excited state of PMI is reduced by NI^-. However, the transient kinetics and spectra indicate that the electron does not proceed all the way back to the porphyrin. We suspect that this is due to

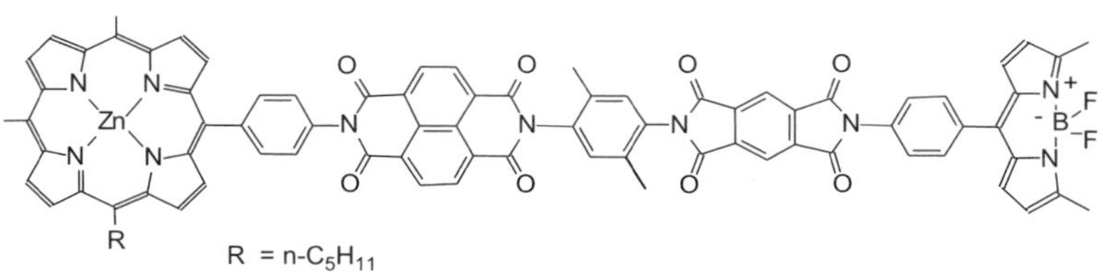

D$_1$ A$_1$ A$_2$ D$_2$

ZnP   NI   PI   DPM

1

R = n-C$_5$H$_{11}$

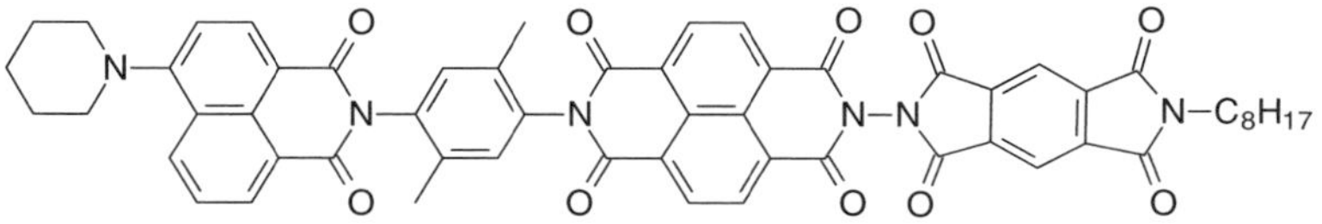

ANI   NI   PI

2

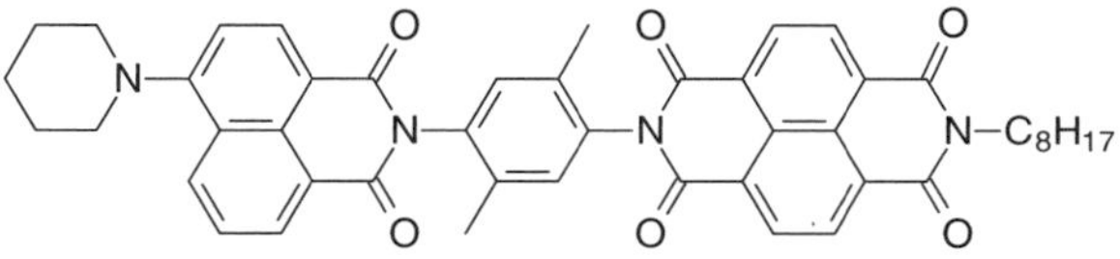

ANI   NI

3

R = n-C$_5$H$_{11}$

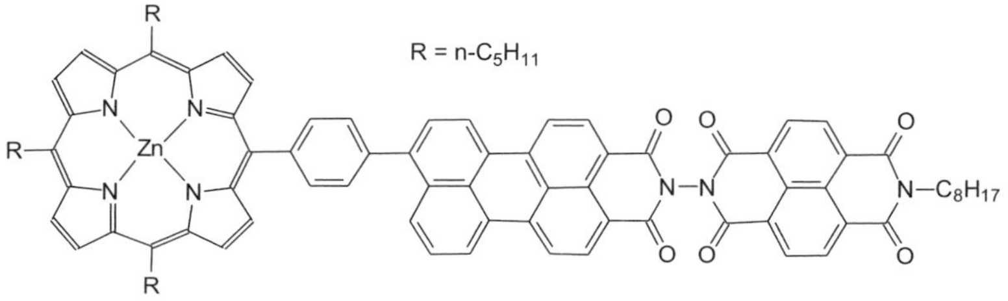

ZnP   PMI   NI
Donor   Bridge   Acceptor

4

the relative electronic coupling matrix elements for the two processes and can be further adjusted by appropriate changes in molecular structure. This experiment suggests that the application of a single photon can be used initiate a relatively long-lived charge separation, while a second photon can be used to change the electronic structure of the bridging molecule resulting in an extremely rapid "turn off" of the charge separation. The charge separation can then be viewed as the storage of information in the molecular system. The diverse, distinct, intense optical absorptions of the intermediates serve as indicators of the "state" of the molecule.

CONCLUSIONS

Our research is exploring the fundamental structural and electronic requirements for ultrafast optical gating of electron transfer in extended arrays of donor-acceptor molecules. We have developed three approaches to a molecular switch that make use of high quantum yield, picosecond time scale electron transfer processes to control the movement of electrons within molecules with the ultimate goal of developing new opto-electronic materials for data manipulation. These approaches include the use of photogenerated electric fields to control electron transfer rates, direct photonic control of ion pair intermediates, and bridging group photo-activation.

ACKNOWLEDGMENT

This research was supported by the Office of Computational and Technological Programs, Division of Advanced Energy Projects, U.S. Department of Energy under contract W-31-109-Eng-38 and the National Science Foundation (CHE-9732840).

REFERENCES

Alcacer, L. (1987). Conducting Polymers: Special Applications. Reidel, Dordrecht.
Asakawa, M., Iqbal, S., Stoddart, J.F. and Tinker, N.D. (1996). Prototype of an optically responsive molecular switch based on pseudorotaxane. *Angew. Chem., Int. Ed. Engl.*, **35**, 976–978.
Aviram, A., Joachim, C. and Pomerantz, M. (1988). Evidence of switching and rectification by a single molecule effected with a scanning tunneling microscope. *Chem. Phys. Lett.*, **146**, 490–5.
Aviram, A. and Ratner, M.A. (1974). Molecular rectifiers. *Chem. Phys. Lett.*, **29**, 277–83.
Baeuerle, P., Gaudl, K.U., Wuerthner, F., Sariciftci, N.S., Neugebauer, H., Mehring, M., Zhong, C. and Doblhofer, K. (1990). Thiophenes. 3. Synthesis and properties of carboxy-functionalized poly(3-alkylthienylenes). *Adv. Mater.* (Weinheim, Fed. Repub. Ger.), **2**, 490–4.

Birge, R.R. (1995). Protein-based computers. *Sci. Am.*, **272**, 90–5.

Blanchard-Desce, M., Marder, S.R. and Barzoukas, M. (1996). Supramolecular chemistry for quadratic nonlinear optics. In *Compr. Supramol. Chem.*, vol. 10 (ed. D.N. Reinhoudt), pp. 833–863. Elsevier, Oxford, UK, Ecole Normale Superieure, Paris, Fr.

Broo, A. (1993). Electron transport through organic molecules with applications to molecular devices. *NATO ASI Ser., Ser. E*, **240**, 163–75.

Burroughes, J.H., Bradley, D.C.C., Brown, A.R., Marks, R.N., Mackay, K., Friend, R.H., Burns, P.L. and Holmes, A.B. (1990). Light-emitting diodes based on conjugated polymers. *Nature*, **347**, 539–541.

Carter, F.L. (1982). *Molecular Electronic Devices I*. Marcel Dekker, New York.

Carter, F.L. (1987). *Molecular Electronics II*. Marcel Dekker, New York.

Carter, F.L., Siatkowski, R.E. and Wohltjen, H. (1988). *Molecular Electronic Devices*. North Holland, Amsterdam.

Carter, T.P., Braeuchle, C., Lee, V.Y., Manavi, M. and Moerner, W.E. (1987). Photon-gated spectral hole burning by donor-acceptor electron transfer. *Opt. Lett.*, **12**, 370–2.

de Silva, A.P. and McCoy, C.P. (1994). Switchable photonic molecules in information technology. *Chem. Ind.* (London), 992–6.

Debreczeny, M.P., Svec, W.A., Marsh, E.M. and Wasielewski, M.R. (1996a). Femtosecond Optical Control of Charge Shift within Electron Donor-Acceptor Arrays: An Approach to Molecular Switches. *J. Am. Chem. Soc.*, **118**, 8174–8175.

Debreczeny, M.P., Svec, W.A. and Wasielewski, M.R. (1996b). A molecular probe of the electric field produced by a photogenerated ion pair. *New J. Chem.*, **20**, 815–828.

DelMedico, A., Fielder, S.S., Lever, A.B.P. and Pietro, W.J. (1995). Rational Design of a Light-Driven Molecular Switch Incorporating an Alizarin-Ru(bpy)2 Fragment. *Inorg. Chem.*, **34**, 1507–13.

Dietz, T.M., Stallman, B.J., Kwan, W.S.V., Penneau, J.F. and Miller, L.L. (1990). Soluble oligoimide molecular lines which have persistent poly(anion radicals) and poly(dianions). *J. Chem. Soc., Chem. Commun.*, 367–9.

Eriksen, J. (1988). *In-situ* Generated Intermediates. In *Photoinduced Electron Transfer*, vol. A (ed. M.A. Fox and M. Chanon), pp. 391–408. Elsevier, Amsterdam.

Feldstein, M.J., Vohringer, P., Wang, W. and Scherer, N.F. (1996). Femtosecond Optical Spectroscopy and Scanning Probe Microscopy. *J. Phys. Chem.*, **100**, 4739–4748.

Feringa, B.L., Huck, N., P.M. and Schoevaars, A.M. (1996). Chiroptical molecular switches. *Adv. Mater.* (Weinheim, Ger.), **8**, 681–684.

Fouquey, C., Lehn, J.M. and Levelut, A.M. (1990). Molecular recognition directed self-assembly of supramolecular liquid crystalline polymers from complementary chiral components. *Adv. Mater.* (Weinheim, Fed. Repub. Ger.), **2**, 254–7.

Gaines, G.L., III, O'Neil, M.P., Svec, W.A., Niemczyk, M.P. and Wasielewski, M.R. (1991). Photoinduced electron transfer in the solid state: rate vs. free energy dependence in fixed-distance porphyrin-acceptor molecules. *J. Am. Chem. Soc.*, **113**, 719–21.

Galoppini, E. and Fox, M.A. (1996). Effect of the Electric Field Generated by the Helix Dipole on Photoinduced Intramolecular Electron Transfer in Dichromophoric .alpha.-Helical Peptides. *J. Am. Chem. Soc.*, **118**, 2299–300.

Garnier, F., Horowitz, G., Peng, X. and Fichou, D. (1990). An all-organic "soft" thin-film transistor with very high carrier mobility. *Adv. Mater.* (Weinheim, Fed. Repub. Ger.), **2**, 592–4.

Gomez-Lopez, M., Preece, J.A. and Stoddart, J.F. (1996). The art and science of self-assembling molecular machines. *Nanotechnology*, **7**, 183–192.

Gosztola, D., Yamada, H. and Wasielewski, M.R. (1995). Electric Field Effects of Photogenerated Ion Pairs on Nearby Molecules: A Model for the Carotenoid Band Shift in Photosynthesis. *J. Am. Chem. Soc.*, **117**, 2041–8.

Gouterman, M. (1978). In *The Porphyrins*, vol. 3 (ed. D. Dolphin), pp. 1–165. Academic Press, New York.

Greenfield, S.R., Svec, W.A., Gosztola, D. and Wasielewski, M.R. (1996). Multistep Photochemical Charge Separation in Rod-like Molecules Based on Aromatic Imides and Diimides. *J. Am. Chem. Soc.*, **118**, 6767–6777.

Gust, D., Moore, T.A. and Moore, A.L. (1993). Molecular mimicry of photosynthetic energy and electron transfer. *Acc. Chem. Res.*, **26**, 198–205.

Haddon, R.C. and Lamola, A.A. (1985). The molecular electronic device and the biochip computer: present status. *Proc. Natl. Acad. Sci. USA*, **82**, 1874–78.

Higgins, D.A., Vanden Bout, D.A., Kerimo, J. and Barbara, P.F. (1996). Polarization-Modulation Near-Field Scanning Optical Microscopy of Mesostructured Materials. *J. Phys. Chem.*, **100**, 13794–13803.

Hopfield, J.J., Onuchic, J.N. and Beratan, D.N. (1988). A molecular shift register based on electron transfer. *Science*, **241**, 817–20.

Jortner, J. (1976). Temperature dependent activation energy for electron transfer between biological molecules. *J. Chem. Phys.*, **64**, 4860–4867.

Kemp, M., Roitberg, A., Mujica, V., Wanta, T. and Ratner, M.A. (1996). Molecular wires: extended coupling and disorder effects. *J. Phys. Chem.*, **100**, 8349–55.

Kuhn, H. (1985). Information, Electron and Energy Transfer in Surface Layers. *Pure and Appl. Chem.*, **53**, 2105–2122.

Kurreck, H. and Huber, M. (1995). Model reactions for photosynthesis — photoinduced charge and energy transfer between covalently linked porphyrin and quinone units. *Angew. Chem., Int. Ed. Engl.*, **34**, 849–66.

Lehn, J.M. (1988). Supramolecular chemistry — scope and perspectives. Molecules-super-molecules-molecular devices. Nobel lecture, 8 December 1987. *Chem. Scr.*, **28**, 237–62.

Lieberman, K., Harush, S., Lewis, A. and Kopelman, R. (1990). A light source smaller than the optical wavelength. *Science*, **247**, 59–61.

Marcus, R.A. (1965). On the Theory of Electron Transfer Reactions. VI. Unified Treatment for Homogeneous and Electrode Reactions. *J. Chem. Phys.*, **43**, 679–701.

Marsella, M.J., Carroll, P.J. and Swager, T.M. (1994). Conducting Pseudopolyrotaxanes: A Chemoresistive Response via Molecular Recognition. *J. Am. Chem. Soc.*, **116**, 9347–8.

Mehring, M. (1989). Concepts of Molecular Information Storage. In *Springer Series in Solid-State Sciences 91: Electronic Properties of Conjugated Polymers III: Basic Models and Applications. Proceedings of an International Winter School, Kirchberg, Tirol, March 11–18, 1989* (ed. H. Kuzmany, M. Mehring, S. Roth and Editors), pp. 484. Springer-Verlag, Berlin, Fed. Rep. Ger.

Metzger, R.M. (1994). The quest for D-σ-A unimolecular rectifiers and related topics in molecular electronics. *Adv. Chem. Ser.*, **240**, 81–129.

Miller, J.S. (1990a). Molecular Materials II. Part A. Molecular Electronics? *Adv. Mater.*, **2**, 378–79.

Miller, J.S. (1990b). Molecular Materials II. Part C. Molecular Electronics? *Adv. Mater.*, **2**, 601–3.

Miller, J.S. (1990c). Molecular materials. II. Part B. Molecular electronics? *Adv. Mater.* (Weinheim, Fed. Repub. Ger.), **2**, 495–7.

Mirkin, C.A. and Ratner, M.A. (1992). Molecular electronics. *Annu. Rev. Phys. Chem.*, **43**, 719–54.

Moerner, W.E., Plakhotnik, T., Irngartinger, T., Croci, M., Palm, V. and Wild, U.P. (1994). Optical Probing of Single Molecules of Terrylene in a Shpol'kii Matrix: A Two-State Single-Molecule Switch. *J. Phys. Chem.*, **98**, 7382–9.

Osuka, A. (1993). Synthesis and photoexcited-state dynamics of aromatic group-bridged carotenoid-porphyrin dyads and carotenoid-porphyrin-pyromellitimide triads. *J. Am. Chem. Soc.*, **115**, 9439–9452.

Osuka, A., Nagata, T., Maruyama, K., Mataga, N., Asahi, T., Yamazaki, I. and Nishimura, Y. (1991). Intramolecular photoinduced electron transfer in fixed distance triads consisting of free-base porphyrin, zinc porphyrin, and electron acceptor. *Chem. Phys. Lett.*, **185**, 88–94.

Parthenopoulos, D.A. and Rentzepis, P.M. (1989). Three-dimensional optical storage memory. *Science* (Washington, D.C., 1883), **245**, 843–5.

Patil, A.O., Heeger, A.J. and Wudl, F. (1988). Optical properties of conducting polymers. *Chem. Rev.*, **88**, 183–200.

Prasad, P.N. and Reinhardt, B.A. (1990). Is there a role for organic materials chemistry in nonlinear optics and photonics? *Chem. Mater.*, **2**, 660–9.

Reimers, J.R., Lue, T.X., Crossley, M.J. and Hush, N.S. (1996). Molecular electronic properties of fused rigid porphyrin-oligomer molecular wires. *Chem. Phys. Lett.*, **256**, 353–359.

Reinhoudt, D.N. and Sudhoelter, E.J.R. (1990). The transduction of host-guest interactions into electronic signals by molecular systems. *Adv. Mater.* (Weinheim, Fed. Repub. Ger.), **2**, 23–32.

Sambles, J.R. and Martin, A.S. (1993). Molecular rectification. *Phys. Scr., T.* **T49B**, 718–20.

Scaiano, J.C., Johnston, L.J., McGimpsey, W.G. and Weir, D. (1988). Photochemistry of organic reaction intermediates: novel reaction paths induced by two-photon laser excitation. *Acc. Chem. Res.*, **21**, 22–9.

Schumm, J.S., Pearson, D.L. and Tour, J.M. (1994). Synthesis of linear conjugated oligomers with an iterative divergent/convergent method to the doubling of monomer purity: rapid access to a 128-Å length potentially conductive molecular wire. *Angew. Chem.*, **106**, 1445–8.

Shimidzu, T., Segawa, H., Wu, F. and Nakayama, N. (1995). Approaches to conducting polymer devices with nanostructures: photoelectrochemical function of one-dimensional and two-dimensional porphyrin polymers with oligothienyl molecular wire. *J. Photochem. Photobiol., A*, **92**, 121–7.

Swager, T.M. and Marsella, M.J. (1994). Molecular recognition and chemoresistive materials. *Adv. Mater.* (Weinheim, Ger.), **6**, 595–7.

Tolbert, L.M. (1992). Solitons in a box: the organic chemistry of electrically conducting polyenes. *Acc. Chem. Res.*, **25**, 561–8.

Tour, J.M. (1996). Conjugated macromolecules of precise length and constitution. Organic synthesis for the construction of nanoarchitectures. *Chem. Rev.* (Washington, D.C.), **96**, 537–53.

Tour, J.M., Wu, R. and Schumm, J.S. (1991). Extended orthogonally fused conducting oligomers for molecular electronic devices. *J. Am. Chem. Soc.*, **113**, 7064–6.

Viehbeck, A., Goldberg, M.J. and Kovac, C.A. (1990). Electrochemical properties of polyimides and related imide compounds. *J. Electrochem. Soc.*, **137**, 1460–6.

Wagner, R.W. and Lindsey, J.S. (1994). A molecular photonic wire. *J. Am. Chem. Soc.*, **116**, 9759–60.

Waldeck, D.H. and Beratan, D.N. (1993). Molecular electronics: observation of molecular rectification. *Science* (Washington, D.C., 1883), **261**, 576–7.

Wasielewski, M.R. (1992). Photoinduced electron transfer in supramolecular systems for artificial photosynthesis. *Chem. Rev.*, **92**, 435–61.

Wasielewski, M.R., Johnson, D.G., Svec, W.A., Kersey, K.M. and Minsek, D.W. (1988). Achieving high quantum yield charge separation in porphyrin-containing donor-acceptor molecules at 10 K. *J. Am. Chem. Soc.*, **110**, 7219–21.

Wiederrecht, G.P., Niemczyk, M.P., Svec, W.A. and Wasielewski, M.R. (1996). Ultrafast Photoinduced Electron Transfer in a Chlorophyll-Based Triad: Vibrationally Hot Ion Pair Intermediates and Dynamic Solvent Effects. *J. Am. Chem. Soc.*, **118**, 81–8.

Willner, I. and Willner, B. (1997). Molecular optoelectronic systems. *Adv. Mater.* (Weinheim, Ger.), **9**, 351–355.

Wilson, E.G. (1995). Nano-molecular electronics: ideas and experiments. *Jpn. J. Appl. Phys., Part 1*, **34**, 3775–81.

Yu, L., Chen, M. and Dalton, L.R. (1990). Ladder polymers: recent developments in syntheses, characterization, and potential applications as electronic and optical materials. *Chem. Mater.*, **2**, 649–59.

14. NEAR FIELD OPTICS: PRINCIPLES AND APPLICATIONS TO NANO MATERIALS

RUGGERO MICHELETTO[a], KEN NAKAJIMA[b],
MASAHIKO HARA[b], WOLFGANG KNOLL[c] and
HIROYUKI SASABE[d]

[a]*Graduate School of Engineering, Dept. of Material Chemistry Kyoto University, Yoshida, Sakyo-ku, 606-8501, Japan*
[b]*Frontier Research Program, The Institute of Physical and Chemical Research (RIKEN), Saitama 351-0198, Japan*
[c]*Max-Plank Institute for Polymer Research, Mainz, Germany*
[d]*Dept. of Photonics Material Science, Chitose Institute of Science & Technology 758-65 Bibi, Chitose, Hokkaido 066-8655, Japan*

INTRODUCTION

The Scanning Near-field Optical Microscope, named as SNOM or PSTM (from Photon Scanning Tunneling Microscope) brought a new powerful tool in the field of imaging and surface science. The ability to investigate the optical properties of the sample with higher resolution than ever before, stimulated a growing interest on these instruments. Comparing with other common techniques like STM or AFM, SNOM has less limitations in the kind of sample. It can be conducting or not conducting, transparent or opaque, soft or hard, thin or thick, in air, vacuum or immersed in liquid. There is not a direct mechanical interaction between probe and sample. The method is purely non contact with all the advantages that this fact implies.

The SNOM can be considered, simply speaking, as the optical equivalent of STM. Instead of electronic current, there is a flux of photons tunneling between the sample and a sharp transparent probe. The optical interaction between tip and sample is rather complex and not yet theoretically solved, however it is generally accepted that the intensity of collected optical field decreases dramatically with sample-probe separation giving high sensitivity to the system.

We will review the principles of SNOM processes giving also a theoretical insight of the state of the research. Then, we will discuss the most interesting applications that have been realized up to now. Some of the author's activities in this field are also presented in detail with the related experimental results.

GENERAL CONCEPTS OF SNOM OPERATIONS

An optical fiber is used as a probing element. This may be coated with metal
to screen the external light. Three typical configurations are shown schematically in Figure 1.

Concerning the first (Figure 1(a), collection mode), the light is impinged
in a glass in total internal reflection condition. Then it is reflected entirely
without any component transmitted on the sample side. However, an optical
field that does not transport energy is still existing, the so called *evanescent
field*. In the presence of a glass probe coupled with it, some photons will
be collected and transported to a detector through the optical fiber.
The intensity of the light is exponentially decreasing with the sample
probe separation. In the hypothesis of an optical homogeneous sample,
we will have an extremely precise morphological information. In this
mode the polarization of the incident light can be freely adjusted, allowing
studies about the properties of the near field in relation with the polarization
vector.

In the so called illumination mode, the light is guided in the optical fiber
and it interacts with the sample at its sharp apex. The scattered light is
collected by an optical detector that can be located laterally near the probe
(i-mode, scattering) or on the opposite side of the sample (i-mode, transmission) as shown in Figure 1(b). In the latter condition, the light will pass
through the sample. Thus, only the experiment on transparent samples will
be possible.

Concerning polarization, it is not possible to set a determined plane of
polarization of the incident wave, because the light loses the condition of
plane wave as it enters the optical fiber. The contrast and transmitted power
appear to be rather high in this mode than in others.

In the third configuration known as reflection mode the sample is illu-
minated directly by the fiber tip that acts simultaneously as a probe for the
scattered light. In other words the light interacting with the sample is partially reflected and collected back by the fiber itself (Figure 1(c)). This
configuration is less popular among researchers due to the experimental
difficulties. However, once the technical problems will be overcome, it could
be the best configuration of all. In fact in this mode the light does not pass
through the sample. Local thickness, the optical path run by the light in the
substrate and other local properties that are not morphologically interesting, will no longer affect the measurement as in other configuration modes.
This allows a more precise image detection, free of spurious information or
artifacts.

As in the STM, the separation between sample and tip is rather small and
must be controlled in order to obtain sharper images. To perform this task
an electronic circuit monitors the light intensity signal and locks it to a

(a)

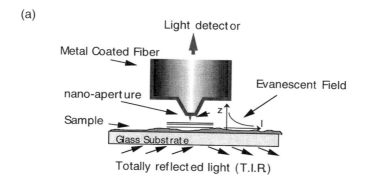

(b)

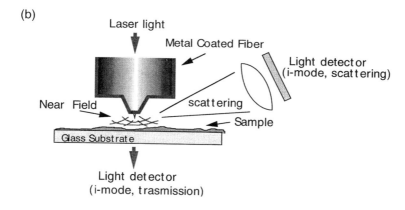

(c)

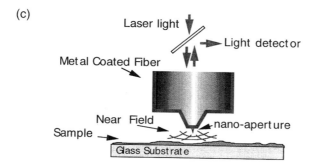

Figure 1 The three main configuration modes for a SNOM system. (a) Collection, (b) illumination and (c) reflection. Two common positioning of the optical detector are shown in the illumination mode (i-mode scattering and i-mode transmission).

determined value by a feedback to the piezo stage. This configuration is commonly called "Optical Feedback".

A different method developed by Betzig *et al.* (Betzig, 1992) introduces a shear force system to perform the same task. The fiber is vibrated at high frequency (order of $50\,KH_z$ or more) by a piezo actuator. A non-contact detector, usually optical, monitors the amplitude of the fiber oscillation. When the fiber approaches the sample surface a reduction of the oscillation amplitude will occur. This is due to proximity interactions between sample and probe. This perturbation is of rather short range (10–100 nm) and it is used as reference signal for a feed-back distance control.

Experimentally, the quality of the images detected by the various SNOM-like devices is strongly dependent on the properties of the tip as, for example, the aperture size and sharpness. Tip production is therefore very important in this field, and we will present here some techniques commonly used and some of the principles of more recent methods.

THEORETICAL ASPECTS

The fundamental processes of all near-field systems could be described as in Figure 2. A light plane wave is incident on a medium interface in condition of total internal reflection. Evanescent field is present on the opposite side and once coupled to a propagating medium (the probe tip) will produce a new real photon that can be detected. Figure 2(a) includes a schematic representation of a SNOM probe-sample area and the corresponding one-dimensional equation for the wave field. The scattering process that occurs in the illumination mode SNOM systems is shown in Figure 2(b). This schematic representation is intended to give an idea of the mechanism involved. The complete three-dimensional solution of Maxwell equation for such systems, characterized by disconnected boundary conditions, might be very difficult to obtain and they are still challenging the theoreticians in the scientific community.

Recent progress in the near field optical microscopy (NOM) increased resolution far over the diffraction limit. This is not the only achievement of this branch of science, in fact the phenomenon itself shed light on the short-range interaction of matter and electromagnetic (EM) field. Fourier optics can be used to demonstrate easily that the diffraction limit to resolution in optical microscopy is not fundamental, but arises on the assumption that the detector (for example a lens) is located several wavelengths away from the sample. If we consider a small detector (or source) scanning in close proximity of the sample, we can obtain an image at resolution depending only by the probe size and probe-to-sample separation, each of which, in principle, can be made smaller than the wavelength of light (Betzig, 1992).

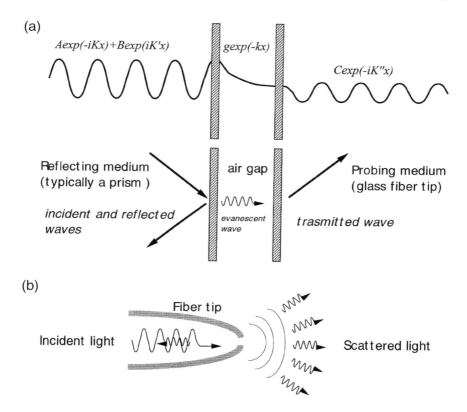

Figure 2   (a) Schematic of the mechanism of photon tunneling in a collection mode SNOM. Basic monodimensional theoretical model is indicated on top of the picture. (b) picture showing the processes involved in the illumination mode configuration.

This fact has been confirmed by a number of experimental results up to now, however they require an appropriate theoretical interpretation. The straightforward analysis of the image obtained is not always leading to a correct understanding of the morphological structure of the sample. For example several authors (Hecht, 1997; Sandogar, 1997) showed that SNOM systems may produce images that represent the path of the probe rather than optical properties of the sample. Thus, a proper understanding of the mechanism involved in the scanning near-field processes is necessary to avoid artifact. However, modeling an entire NOM system is rather complex and can be met only by extended numerical simulations (Novotony, 1994).

Besides, there are many different configuration modes, which could be generally classified as illumination, collection, reflection, and each of them should be modeled in a completely different way. In this respect, recent publications have attempted to understand the process involved. However, the explanations offered are in the approximate form and a generally accepted rigorous theory is not yet established (Zvyagin, 1997).

CURRENT EXPERIMENTS

The first experiments about scanning near field optics were performed by Betzig et al. in the late eighties. He firstly reported sub-wavelength resolution images with an optical scanning device. Betzig demonstrated that the resolution power is limited in principle by the aperture size of the probing element and not by the wavelength. His images, obtained purely by optics, showed for the first time the breaking of diffraction limit by near field and excited a number of researchers all over the world. His group introduced also a sample-probe distance controller based on the shear force interaction between sample and probe. This idea was rather well accepted and now is used in several of apparatus developed by different authors.

Shear force control system consists of a probe that is vibrated at a high frequency by a piezo actuator. A laser light is diffracted by the probe tip in order to create a pattern on an optical detector, typically a photo diode. The amplitude of vibration is monitored by detecting the intensity of these diffraction fringes. When the probe approaches the surface of the sample there is an interaction and the amplitude of vibration is reduced. This can be used as a feedback signal to control the position of the tip (see Figure 3). Even if this control system does not provide optical information, it is often used in conjunction with SNOM systems because it is easily applicable to a glass fiber probe and the resolution (around 10 nm) and stability are appropriate.

Other researchers started investigations in this field. Courjon, Phol and Reddick (Pohl, 1988; Courjon, 1989; Reddick, 1989) were the first groups developing Near-field Optical Microscopes. One of the most striking properties of near field optics is the fast exponential decay of evanescent field that give rise to extremely high resolution in the vertical direction. One of us (RM) developed a near-field scanning microscope, diffraction limited laterally, that could resolve better than 1Å in the direction normal to the plane of the sample (De Marco, 1993). This system works in reflection mode, i.e. the light does not traverse the sample. It can be strongly absorbent, even a metal. The process of reflection and tunneling will always allow the mapping of the sample. This instrument has an extremely high optical sensitivity that give rise to a vertical resolution of less than 0.5 Å. However the scanning is performed without the aid of a sharp probe, thus the lateral resolution is diffraction limited. A schematic of this particular system is shown in Figure 4.

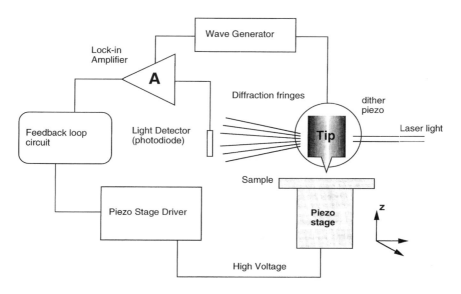

Figure 3 The scheme of a Shear force control system.

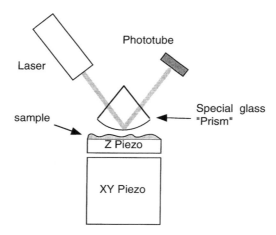

Figure 4 The core of a high sensitivity reflection system. The reflecting element is a special "prism" with a spherical face. This SNOM has an extremely high vertical resolution (0.5 Å), but it is diffraction limited laterally.

High resolution imaging in water has been obtained by different groups. SNOM is a purely non contact system hence the probing element can be fully immersed in water without problems. Ohtsu and other groups developed optical systems that performed at extremely high resolution even with samples in liquid. The flagellars of a biological specimen of diameter around 25 nm were observed in liquid in collection mode (Naya, 1997). Similar systems can be applied in the investigation of living biosamples or other kind of experiments.

Near field Optics has been applied not only to imaging experiments, but also to experiments of general physics. Saiki *et al.* reported the isolation of a single quantum dot observed directly with a SNOM-like system. Other authors are investigating fluorescence, surface plasmon resonance on surfaces, Raman spectroscopy and other various phenomena, taking the advantage of the nano-localized near-field realized by optical probe (Ferrel, 1992; Coello, 1998; Tarrach, 1995).

The authors developed a SNOM-Shear Force integrated system. The apparatus is based on the conventional principles of near-field, however, it is designed in an extremely flexible way. Our system is capable to operate in all the optical modes, illumination, collection and reflection. We can easily modify the angle of incidence of the light impinging on the sample, allowing the excitation of surface plasmon on a metal interface. In this way we can operate our instrument as a SPOM (surface plasmon optical microscope). With straightforward modification of the optical set up it is possible to pass from an optical mode to the other. The data acquisition is entirely designed by us, and we can accept up to 8 input channels. Our home-made software can detect and display three different signals recorded simultaneously in the same scan. We can produce a morphological map with the Shear Force data together with two different images from other signals. Typically one of the signal is coming from the fiber probe (i.e. in collection mode). Another one could be a processed version of the same signal coming from the fiber, for example through a lock-in amplifier locked to the shearing frequency as shown in Figure 5, or any other interesting data produced by the scanning.

The analysis of the images detected can give interesting insight on the conformation and optical properties of the sample. In Figure 6 we show two simultaneous scans of a CD-ROM surface to demonstrate high resolution in both shear-force and illumination mode. We are going to present the details of this new developed apparatus in a forthcoming paper.

TIP FABRICATION

Fibers are sharpened by using the difference in etching rates between core and cladding in a selective etching solution of buffered HF acid. Advantage of high reproducibility, mass production, ability to tailor shapes of resulting fibers by controlling etching conditions make this method very reliable. The fabrication procedure was originally developed by Ohtsu (Ohtsu, 1995).

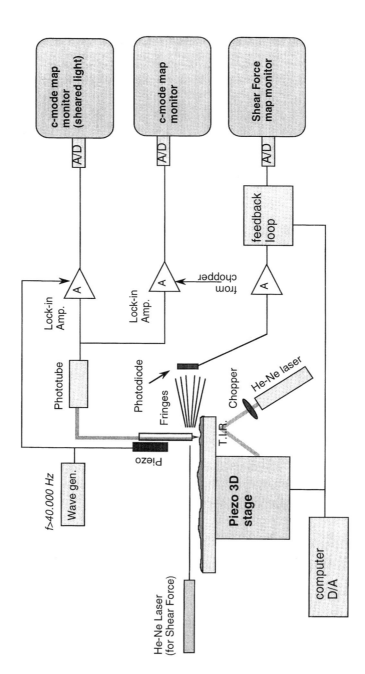

Figure 5   The scheme of author's SNOM/Shear force integrated system.

R. MICHELETTO *et al.*

(a)   (b)

Light direction

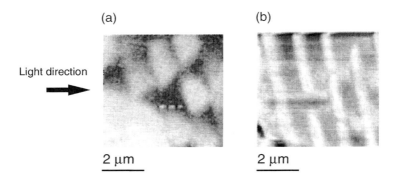

2 μm   2 μm

Figure 6 (a) SNOM collection mode map of CD-ROM dots. Dots height is 120 nm and the distance between them is 1.25 μm. (b) Shear Force map of the same sample. The two images are taken simultaneously.

We operated some modifications to this method in order to obtain the desirable tip shape for SNOM and SNOM/STM imaging. A glass fiber tip (125 μm diameter, 8 μm 25% Ge doped core) is immersed in etching solution. The composition of it is given in volume ratios of NH_4F (40 wt.%): HF (50 wt.%): $H_2O = 2:1:1$. We etched the fiber for 2 hours at room temperature (25 °C). We observed a reduction of the diameter of the cladding in the order of 30−40 μm.

Because of the high concentration of HF components, etching rate is the same for both core and cladding. Thus no tip is created in this process and the top of the fiber looks still flat. However, this first etching is important for the mechanical stability of the tip and for other reasons explained in detail in several papers (Mononobe, 1997, Pangaribuan, 1994). The second etching is performed with the same chemicals at a different volume ratio: NH_4F (40 wt.%): HF (50 wt.%): $H_2O = 10:1:1$. In this case the etching speed of the core becomes lower than the cladding one. After one hour of etching the core will protrude from the rest of the fiber with a very sharp conical shape.

This tips can be used without any other modification in experiment of near-field optics where a great sensitivity is necessary and where the spatial resolution is less important. To improve the performances of the tip, it is possible to coat the whole body of the fiber with metal. The procedure we used is based on the method introduced by Mononobe *et al.* (Mononobe, 1997).

Only in the apex we will create an aperture to allow the transmission of light. This aperture must be as small as possible and will determine the resolution of the images collected. Concerning the metal coating we used an ion sputter with platinum in a low vacuum (about 0.1 Torr) machine. After

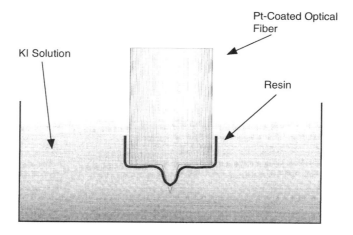

Figure 7 Scheme of showing the procedure to create a nanometric aperture on a metal coated glass fiber tip (see text for details).

Figure 8 A SEM picture of a metal-coated fiber tip fabricated by the authors. A nanometric-sized transparent aperture is clearly visible. Size of the aperture is around 0.1 μm, radius of the apex of the tips is about 10 nm.

about 20 minutes of sputtering the fiber will be covered by 100 nm of metal. To create the small aperture we firstly dip it in an acrylic resin for a few seconds. The resin will cover the whole fiber. Only the sub-micrometer top region of the protruding sharpened core will not be covered. In fact the surface tension will not allow the resin film to follow the sharp profile of the tip (see Figure 7).

In a last chemical etching by KI-I_2 solution, the platinum at the apex of the fiber will be removed (KI:I_2:$H_2O = 20$:1:400). We could obtain tips with rather high reproducibility as shown in Figure 8. These tips can be used directly for high resolution SNOM experiment. A further metallization is necessary to make them conducting. We simply sputter again as described above for a short period of time (5 min). The platinum layer obtained is proved to be thick enough to get electrical contact to the sample in order to perform STM imaging.

Production of these tips requires particular fibers with a Germanium doped core of 23% doping ratio. They are not commercially available, however we succeeded to produce pencil type tips with commercial fibers with a Germanium doped core of 3% doping ratio. The results shown previously (see Figure 6) are obtained with these tips. We are going to present the details of the method of production elsewhere.

NEW APPLICATIONS AND PERSPECTIVES

We will discuss the future developments that are currently foreseen in the field of Near Field Optics. Particular applications are expanding in the direction of imaging, so we will propose the development of new instruments, like the integration of near field system with a conventional STM. We will also mention briefly other new applications in near field optics that are promising for future development.

Integration of SNOM with STM

One of the weak point of SNOM system is considered to be the large and poorly controlled separation between sample and tip. Recently a number of groups have been working in the direction of improving this feature using a STM in cooperation with SNOM (Lieberman, 1993; Garcia, 1994; Inouye, 1995). In such instruments, the distance control feedback is driven directly by the STM permitting a scanning at rather close proximity (in the order of 10 Å). This value is very stable and it is controllable easily by the current setting.

Also, as our group is working on the same task, we believe that the use of our particular chemical etched glass fiber tip will perform well in a STM/ SNOM system. These tips have already shown interesting performances in pure SNOM mode (Naya, 1997). We expect a general improvement in spatial resolution compared with other proposed instruments.

In Figure 9 is shown the scheme of the system that the authors are currently developing. The head of a commercial STM system is replaced by a home-made head that has been designed to host a glass fiber tip. The latter is made conductive by metal coating using a thin layer of platinum that is sputtered on the body of the fiber as described above. The purpose of the

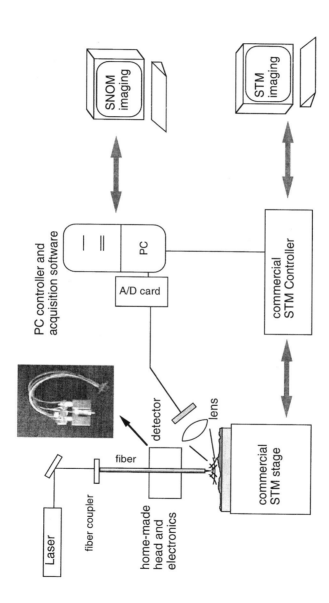

Figure 9 Scheme of author's SNOM/STM integrated system. Inset is the picture of the home made head. This head can host indigenously fabricated metal coated fiber tips. The head is fully compatible with a commercial STM head.

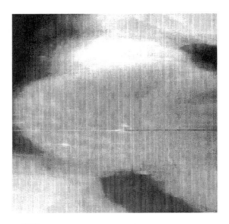

30 nm

Figure 10   Au(111) on mica mapped by the SNOM/STM apparatus developed by the authors.

procedure is double: first of all the glass fiber has to be conductive in order to be used as a conventional STM tip, secondly the light should be detected only at the apex of the tip. The aperture size that we could realize with this method are of the order of 50 nm. We designed the pico-ampere amplifier and installed in our special STM head. We scanned Au(111) sample with the glass tips fabricated by us. Atomic steps were clearly detected demonstrating high resolution performances. The obtained result is shown in Figure 10.

An apparatus of this kind can be applied in several studies taking the advantage of the extreme high localization of the optical field. For example experiments of optical storage are foreseen easily, provided the presence of a photosensitive medium that modify somewhat its optical or morphological properties by a proper illumination. The STM tip can read any structural change in the film in real time and high resolution, allowing detailed investigation on the photochemical process involved.

Surface Plasmon Microscopy

Surface plasmons can be easily excited by the Kretschmann attenuated total internal reflection configuration (Kretschmann, 1963). Surface plasmon resonance is strongly sensitive to the local changes in dielectric properties of the metal-dielectric interface. An interesting application of near-field technology is to produce an optical microscope to detect surfaces in such conditions. Scanning can allow the study of the distribution of chemical and physical structures in a dielectric sample at sub micrometer resolution. Several authors as Kim *et al.* (Kim, 1996) reported the development of similar devices known as SPOM (Scanning Plasmon Optical Microscope).

Atom Manipulation

The strong localization of the field realized by SNOM tips induces an extremely high field gradient that could not be realized when using ordinary propagating light. An atom, which scatters an evanescent photon into a propagating one by resonance absorption and spontaneous emission, gets a large recoil momentum parallel to the probe-tip surface. This acts as a cooling force until the atom momentum and the recoil momentum get close. The atom feels also a strong dipole force gradient in the direction normal to the tip surface.

These two phenomena combined can be utilized to trap and cool atoms. Hori (Hori, 1993) described this method giving a quantitative evaluation of the experimental parameters necessary to realize an actual trapping system. The theoretical idea described by him gives a scenario in which an atom is cooled in the proximity of a nanometric fiber-probe by the cooperation of the two forces briefly described above: the centrifugal recoil due to the spontaneous emission and the dipole attractive force due to field gradient. The balance of these interactions produces a potential valley along the probe-tip surface. The atom will move in it like a pinball trapped in the optical standing wave (see Figure 11).

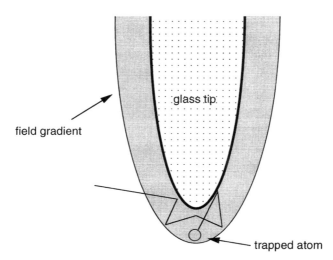

Figure 11 The mechanism of atom trapping by a SNOM tip as proposed by Hori.

This theoretical insight, as far as we know, has not yet been reproduced in an experiment by anyone, however its potential in the field of the possible applications is so high that we believe several groups are working on this problem.

SNOM Measurement in Liquid

For real understanding of biological samples, it is necessary to conduct *in vivo* observation in the natural environment of the specimen. Although it is possible to conduct these experiments with conventional optical microscopes, their resolution is limited by wavelength. It has recently been reported (Mertesdorf, 1997) that a near-field optical microscope will surpass this limit with a typical SNOM system modified for this purpose (see Figure 12 for a typical configuration). This apparatus operates a shear force distance control even with the tip partially immersed in liquid.

Other authors (Naya, 1997), to simplify the system configuration, succeeded to work in liquid without shear force. The sample probe separation

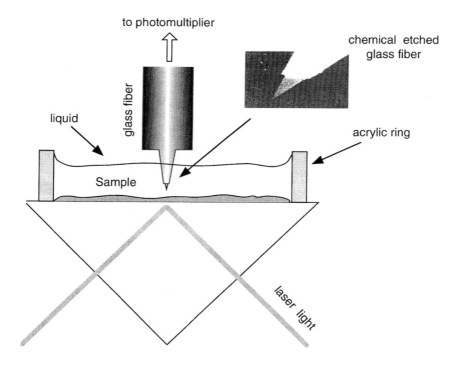

Figure 12   The scheme of a SNOM apparatus working in liquid.

was controlled by a simple optical feedback. In other words, a feedback circuit is monitoring the intensity of the collected light and keeps it constant.

In order to work in liquid, the sample holder usually consists of a Teflon ring in which water or other liquids are inserted (Naya, 1997). There is no need for particular sealing. In fact the surface tension of the liquid itself will avoid leakage in most of the cases.

Other recent reports show other measurements in liquid, as the SNOM/AFM system realized by Keller *et al.* (Keller, 1997), in which the atomic force high resolution characteristics are integrated with near field optics.

Local Phenomena (Single Molecule, Single Atom Emission)

A number of reports from different authors show that localized phenomena may be fully studied by near-field optical techniques. These studies can be performed with high resolution and they are not limited by diffraction. Other methods like STM and AFM may perform better from the spatial resolution point of view, but can not detect optical phenomena and they are quite destructive acting on soft and delicate samples.

Fluorescence imaging of single molecules has been reported recently by several authors, for example Tarrach *et al.* (Tarrach, 1995). The molecule used was a rhodamine-6G (R-6G) which has a maximum absorbance of the excitation light around 520 nm and has a maximum emission at 580 nm.

In the semiconductor science the introduction of SNOM has brought several insights. Measuring photo-luminescence spectral profile from quantum dots in GaAs/AlGaAs single quantum wells has been made possible by the use of near field techniques (Toda, 1996)

A two-dimensional mapping of the light emission of a p-n diode junction has been also reported by Eah *et al.* (Eah, 1997). The mechanism involved in this phenomenon is already known, but high-resolution optical data were taken for the first time in this report.

A number of recent papers use near-field techniques to detect fluorescence spectroscopy on a great variety of samples. For example the labeling of biological samples with fluorescent dye is one of the most common methods. Fluorescence spectral profiles, lifetimes and quantum efficiencies depend strongly on their surroundings, giving much information to the researchers, besides the pure morphological data.

FINAL REMARKS AND ACKNOWLEDGMENT

Near field optics, and particularly SNOM, are recent techniques that are getting increasingly popular. We gave an overview of representative works in the field, also, we presented our own progresses. Examining the status of the newest researches in this area, we understand that the role of integration between SNOM and other probing techniques is fundamental. The non-contact characteristic of SNOM is promising in the study of interface, allowing analysis without abrupt interference with the natural status of the sample. This feature is particularly useful in the study of biological or living samples. Mechanical properties of thin films can be also foreseen, due to the extreme sensitivity of these techniques to small vertical displacements. In spite of more popular methods as STM or AFM, SNOM uses light as a probing medium, thus the wavelength spectroscopy is possible. This can give the researcher a huge amount of new information to investigate with.

A number of optical experiments related with spectroscopy that were performed in bulk conditions, now can be reproduced with SNOM, breaking the diffraction limit. Fluorescence, time resolved spectroscopy, Raman spectroscopy, surface plasmon are newly investigated with near-field. New insight on the nature of these phenomena can be examined and the physical and chemical properties of nano-sized materials can be studied with these techniques. Great progress has been done in the theoretical interpretation of the near-field process. Though, a complete and generally accepted model is still under study. SNOM approach has a lot of potential and is still open to further development.

We want to express our gratitude to Drs T. Isoshima, R. Uma Mahesh-wari, S. Mononobe and Prof. Ohtsu for their kind help and discussion for this manuscript.

REFERENCES

Betzig, E. and Trautman, J.T. (1992). *Science*, **257**, 189.
Betzig, E., Finn P. and Weiner, J.S. (1992). *Appl. Phys. Lett.*, **60**, 20, 2484.
Coello, F.A., Bozhevolnyi, S.I. and Pudonin, (1998). *Proceedings of the SPIE — The International Society for Optical Engineering*, **536**, 3098.
Courjon, D., Sarayeddine K. and Spajer, M. (1989). *Opics Comm.*, **71**, 23.
De Marco, A., Micheletto, R., Trabucco A. and Violino, P. (1993). *Optics Comm.*, **95**, 210.
Ferrell, T.J., Sharp, S.L. and Warmack, R.J. (1992). *Ultramicroscopy*, **42–44A**, 408.
Garcia-Parajo, M., Carminati, E. and Chen, Y. (1994). *Appl. Phys. Lett.*, **65**, 12, 1498.
Hecht, B., Bielefeld, H., Inouye, Y., Pohl, D.W. and Novotny, L. (1997). *J. Appl. Phys.*, **81**, 6, 2499.
Hori, H., Pohl, D.W. and Courjon, D. (1993). eds., *Near Field optics*, **105**.
Inouye, Y. and Kawata, S. (1994). *Journal of Microscopy*, **178**, 1, 14.
Keller, T.H., Rayment, T., Klenerman, D., Stephenson, R.J. (1997). *Review of Scientific Instruments*, **68**, 3, 1448.
Kim, Y-K., Ketterson, J.B. and Morgan, D.J. (1996). *Optics Letters.*, **21**, 3, 165.
Kretschmann, E. and Raether, H. (1963). *Z. Naturforsh*, **23a**, 2135.
Lieberman, K. and Lewis, A. (1993). *Appl. Phys. Lett.*, **62**, 12, 1335.

Mertesdorf, M., Schonhoff, M., Lohr, F. and Kirstein, S. (1997). *Surface and Interface Analysis*, **25**, 10, 755.

Naya, M., Micheletto, R., Mononobe, S., Uma Maheswari, R. and Ohtsu, M. (1997). *Applied Optics.*, **36**, 7, 1681.

Mononobe, S., Naya, M., Saiki, T. and Ohtsu, M. (1997). *Appl. Optics*, **36**, 7, 1496.

Novotny, D., Pohl, D. and Regli, P. (1994). *J. Opt. Soc. Am.*, **11A**, 1768.

Ohtsu, M. (1995). *OPTOELECTRONIC devices and technology*, **10**, 2, 147.

Pangaribuan, T., Jiang, S. and Ohtsu, M. (1994). *Scanning*, **16**, 362.

Pohl, D.W., Fisher, U.Ch. and Duricg, U. (1988). *Proc. SPIE.*, **897**, 84.

Reddick, R.C., Warmack R.J. and Ferril, T.L. (1989). *Phys. Rev. B.*, **39**, 767.

Sandoghdar, V., Wegscheider, S., Krausch, G. and Mlynek, J. (1997). *J. Appl. Phys.*, **81**, 6, 2492.

Sang-Kee Eah, Wonho Jhe, Saiki, T. and Ohtsu, M. (1997). *Optical Review*, **3** 6B, 450.

Tarrach, G., Bopp, M.A., Zeisel D. and Meixner, A.J. (1995). *Rev. Sci. Instrum.*, **66**, 6, 3569.

Toda, Y., Kourogi, M., Ohtsu, M., Nagamune, Y. and Arakawa, Y. (1996). *Applied Physics Letters*, **69**, 6, 827.

Zvyagin, A. and Ohtsu, M. (1997). *Optics Comm.*, **133**, 328.

15. THE CURRENT RESEARCH ACTIVITY OF SPIN INVESTIGATION IN THE NANOELECTRONICS LABORATORY

KOICHI MUKASA*

*Nanoelectronics Laboratory, Graduate School of Engineering,
Hokkaido University, Sapporo 060-8628, CREST "Spin Investigation Team",
Japan Science and Technology Corporation (JST)*

I believe electronics has two trends in the future. That is, nanometer electronics and femto second electronics. From these trends our research will be one: Spin Investigation, Two: Molecular Device and its material made by L-B method and three: Peltier Device and its material.

There are three on-going studies on spin investigation in my laboratory as shown in Figure 1. The first area of study is development of Spin SPM. The Spin SPM delivers a real-space image of the surface spin with atomic resolution. Two instruments are being developed: The first is SP-STM (Spin Polarized STM)[1] and the second, EFM (Exchange Force Microscope).[2] The second area of study is spin engineering of new materials of which the magnetic state is evaluated by Spin SPM on atomic scale. It's so to speak, magnetism on an atomic scale. There are three different aspects of evaluation, the first: spin state, the second: surface magnetism, and the third: surface magnetism of nonmagnetic materials. The first, spin state, which includes the magnetism of fine particles, surface magnetic anisotropy and giant magnetic moment which is a proper characteristic of magnetic material surface, and giant magnetoresistance which is observed in the super lattice structure. The second, surface magnetism, which includes a magnetic recording bit and magneto-optical bit whose linear bit length is 0.1 to 0.2 μm now and is forecasted to be less than 30 nm at beginning of the 21 century. The last, surface magnetism of nonmagnetic materials by EFM, which includes surface spin of organic or bionic materials and chemical reaction in which spin takes part. The last area of study is spin electronics, which is derived from the basic study of the development of Spin SPM. In the study of SP-STM, the itinerant-electron spin is the main player and we have to study spin relaxation phenomena of the itinerant electron. From these studies, we can reach Spin quantum electronic devices in the future, which have low-noise and very high-speed properties. In the study of EFM,

* Tel.: + 81-11-706-6539, Fax: + 81-11-706-7803, E-mail: mukasa@nano.eng.hokudai.ac.jp

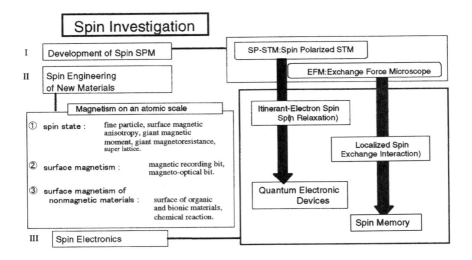

Figure 1

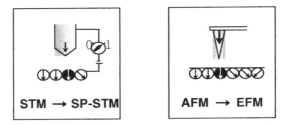

Figure 2

localized spin is the main player and we have to study exchange interaction between spins of sample and tip and it's mechanism. From these studies, we can reach atom spin memory, the ultimate goal of memory which can be achieved, if the localized spin in each atom can be reversed by exchange interaction.

At first we describe Spin Polarized STM, abbreviated SP-STM as shown in Figure 2. The STM delivers a real-spin image of a surface with atomic resolution. By using the additional spin-dependence of the tunneling current, a magnetic image can be obtained. Our SP-STM apparatus should have

1. a well defined tip, especially with respect to the electron spin polariza-
    tion of the tip and
2. a non-magnetic tip, so that a magnetic field from the tip does not affect
    the spin configuration of the sample.

AFM is similar to STM, except that it detects forces rather than currents.
Compared to STM, AFM has the advantage of being able to analyze
insulating objects. When an exchange interaction between a sample and a tip
is measured, we refer to this analysis as exchange force microscopy (EFM).

   One of the requirements of our SP-STM is to have a well defined tip,
especially with respect to electron spin polarization of the tip. We are
planning to combine our SP-STM with three techniques as shown in
Figure 3. Firstly, a δ-doped GaAs structure will be made by MBE,[3] then the
structure will be milled into a tip by Focussed Ion Beam (FIB),[4] and finally
the spin polarization of the tip will be measured by Mott spin detector.[5] All
of these procedures will be carried out in a vacuum.

   By MBE, we have made a δ-doped GaAs/AlGaAs heterostructure with a
spin relaxation time in the order of n sec, which is longer than that of bulk

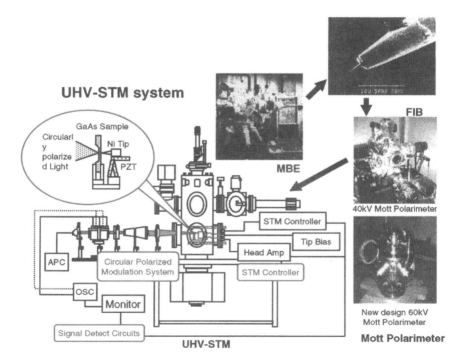

Figure 3

GaAs.[3] Long spin relaxation time would enable the separation of the tunneling part from the light illuminating part on the tip. By FIB, we have developed tips with a very sharp apex of about 10 nm in radius. At present we are trying to fabricate Si and GaAs tips. We have also developed a compact Mott spin detector with an acceleration voltage of 60 kV.

The five papers of our members include "Development of Exchange Force Microscopy" by Koji Nakamura, which is the calculated result of exchange interaction between tip and sample and is the forecast for the experiment of EFM. Makoto Sawamura provides the paper entitled "Design of Molecular Spin Devices — Control of Surface Spin States —" which is the theoretical forecast of future spin devices. Guido Eilers presents the paper entitled "Observation of Surface Magnetism of Magnetic Multilayers", which is the basic study of magnetism and electron conduction in the super lattice structure. Kazuhisa Sueoka discusses "Spin Polarized Scanning Tunneling Microscopy". The paper, "Electron Tunneling into Disordered Materials"[6] by Eiji Hatta, is concerned with elastic and inelastic tunneling in LB thin film.

REFERENCES

1. Sueoka, K. *et al.* (1993). *Jpn. J. Appl. Phys.*, **32**, 2989.
2. Nakamura, K. *et al.* (1997). *Phys. Rev. B*, **56**, 3218; Mukasa, K. *et al.* (1994). *Jpn. J. Appl. Phys.*, **33**, 2692.
3. Kimura, K. *et al.* (1996). *J. Mag. Soc. Japan*, **20**, 253 (in Japanese).
4. Sasaki, M. *et al.* (1995). *Vacuum*, **38**, 285 (in Japanese).
5. Sasaki, M. *et al.* (1994). *J. Mag. Soc. Japan*, **18**, 257 (in Japanese).
6. Hatta, E. *et al.* (1997). *Solid State Com.*, **102**, 437.

16. DEVELOPMENT OF EXCHANGE FORCE MICROSCOPY

K. NAKAMURA[a,*], H. HASEGAWA[b], T. OGUCHI[c], K. SUEOKA[d], K. HAYAKAWA[a] and K. MUKASA[d,e]

[a]Catalysis Research Center, Hokkaido University, Sapporo 060-0811, Japan
[b]Department of Physics, Tokyo Gakugei University, Koganei, Tokyo 184, Japan
[c]Department of Materials Science, Hiroshima University, Higashi-Hiroshima 739, Japan
[d]Nanoelectronics Laboratory, Graduate School of Engineering, Hokkaido University, Sapporo 060-0811, Japan
[e]CREST "Spin Investigation Team", Japan Science and Technology Corporation (JST)

The application of the atomic force microscopy (AFM) measurements to magnetic material has attracted much attention from the scientific and technological point of view. The magnetic force microscopy (MFM) detects the force arising from the long-range interaction between magnetic dipoles of the tip and sample. The typical tip-sample separations in the MFM are of the order of more than 10 nm and the spatial resolution is of the order of 10 nm to 100 nm. It is fairly certain that improvement of the resolution might be made by probing the short-range exchange force rather than the long-range magnetic dipole force.[1-4]

In this papers, we have theoretically investigated the feasibility of the exchange force microscopy (EFM) which probes the short-range exchange force between the tip and sample by carrying out first-principles calculations with the use of the FLAPW method within a local density approximation.

As a model system for the tip and sample, we assume the two three-atomic-layer Fe(100) films which are separated by the distance d, as shown in Figure 1. Surface atoms of the one film are assumed to be facing the hollow sites on the surface of the other film. Because bulk Fe is in the ferromagnetic ground state, magnetic moments in each three-layer film is taken to be in the ferromagnetic alignment. As far as the relative orientation of moments in the two films are concerned, however, we consider the parallel (P) and anti-parallel (AP) configurations in order to calculate the exchange interaction and the exchange force between the two magnetic

* Tel.: +81-11-706-3271, Fax: +81-11-706-4959, E-mail: nakamura@nano.eng.hokudai.ac.jp

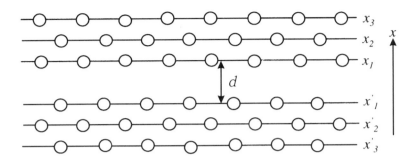

Figure 1 Schematic representation of the two three-atom-layer Fe(100) films adopted in our calculation, surface atoms of the one film facing the hollow sites on the surface of the other film. Open circles represent Fe atoms and the lines denote the layer planes which are referred to as the x_n and x'_n layers ($n = 1-3$).

films. Our LAPW calculations are carried out by changing the film-film separation d from 1.4 Å to 5.0 Å with an assumption that the internal atomic coordinates of each film are rigidly fixed.

Figure 2 shows the spin magnetic moments calculated within the muffin-tin spheres of Fe atoms in the three layers of the upper film, which are referred to as the x_1, x_2 and x_3 layers (Figure 1). The film-film separation d in the abscissa is denoted relative to the lattice constant of bulk Fe ($a = 2.83$ Å). We should note that the separation of $d/a = 0.5$ corresponds just to the interlayer distance of bulk Fe. Due to the symmetry in our model, the magnetic moments in the x'_n layer of the lower film are the same as those in the x_n layer of the upper film in the P configuration, while in the AP configuration those in the x'_n layer has the same magnitude but with the opposite sign as those in the x_n layer. For both P and AP configurations, the magnetic moments at the outer x_3 layer are enhanced to be about 2.9 μ_B, which is almost identical to the surface magnetic moment obtained by slab calculations.[5,6] The magnetic moments at the central x_2 layer are about 2.3 μ_B, which is close to the experimental bulk value and to calculated one at the central layer of the slab.[6] The magnetic moments of the x_2 and x_3 layers for both P and AP configurations are almost independent of the film-film separation. On the contrary, magnetic moments on the x_1 layer change considerably at $d/a < 1$, where the magnetic moments reduce significantly from the surface value to the bulk one as decreasing d. Nevertheless, the change in the magnitude of moments of the x_1 layer becomes insignificant for $d/a > 1$. We notice that the magnetic moments for the AP configuration are always smaller than that for the P configuration.

Figure 3 shows the total energies, E_P and E_{AP}, for the P and AP configurations as a function of d/a. When the films are moved from the bulk

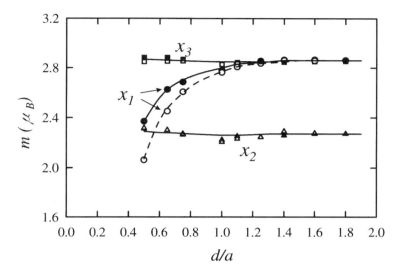

Figure 2 Magnetic moments in the muffin-tin sphere of an Fe atom in the x_1 (circles), x_2 (triangles) and x_3 (squares) layers as a function of the film-film separation d normalized by the lattice constant of bulk Fe ($a = 2.83 \text{ Å}$). Solid and open marks stand for the P and the AP configurations, respectively.

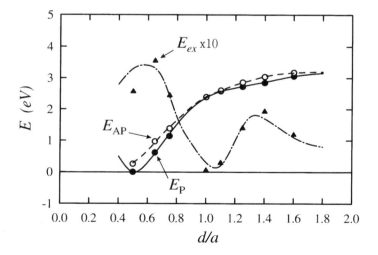

Figure 3 The film-film separation dependence of total energies in the P (E_P: filled circles) and AF (E_{AF}: open circles) configurations, and of the exchange energy defined by $E_{ex} = E_{AF} - E_P$ (triangles). The reference energy for E_P and E_{AF} is the total energy for the P configuration at $d/a = 0.5$.

position at $d/a = 0.5$, the total energies for both E_P and E_{AP} increase. The exchange interaction energy, E_{ex}, defined by $E_{ex} = E_{AP} - E_P$ is also plotted in Figure 3. In the all region ($0.5 < d/a < 1.7$) investigated, we get the positive E_{ex}, which means that the P configuration is more favorable than the AP one. This is expected from the fact that the resulting magnetic configuration corresponds to a natural stacking of ferromagnetic Fe in bulk. The exchange interaction energy has an RKKY-type oscillation with a period of about $d/a = 0.7$ ($\sim 2\,\text{Å}$). The peak values of E_{ex} are $0.35\,\text{eV}$ at $d/a = 0.6$ and $0.19\,\text{eV}$ at $d/a = 1.35$. In the case of $d/a < 1$, where the relevant d orbitals must have a large overlap, the exchange force may arise from the direct exchange couplings between d states of the two films. In the case of $d/a > 1$, on the other hand, the exchange coupling is expected to be mediated through delocalized s and p electrons, where the perturbation of the approaching ferromagnetic film to magnetic moments in the other film is negligibly small as mentioned in Figure 2.

Figure 4 shows the forces acting on the upper film for the P and AF configurations. The force direction is perpendicular to the film surface. The exchange force defined by $F_{ex} = F_{AP} - F_P$ has an oscillation against d/a with the period of about 0.7, which is identical with the period of the exchange interaction (Figure 3). The magnitude of the calculated exchange force is of the order of 10^{-9} at $d/a < 1$ while it is of the order of $10^{-10}\,\text{N}$ at $1 < d/a < 1.5$. These forces are large enough to be observed because the

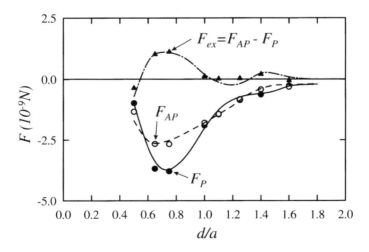

Figure 4 The film-film separation dependence of total forces in the P (F_P: filled circles) and AF (F_{AF}: open circles) configurations, and of the exchange force defined by $F_{ex} = F_{AF} - F_P$ (triangles).

sensitivity of the current AFM is of the order of 10^{-12} to 10^{-13} N,[7] although it would be necessary to take into account an actual shape of the apex tip before we deduce definite conclusions. A first-principles calculation adopting more realistic tips is under consideration.

In actual measurements of the exchange force by using the EFM, we would face technical difficulties such as the snap-in effect: when the tip-sample distance comes in the point of force instability where the gradient of the force exceeds the magnitude of the lever stiffness, the tip and surface would snap into contact. In order to measure the exchange force, we require an exploration of the way to avoid the snap-in effect. Recently, a continuous measurement of the force curve between tip and sample, as approaching towards contact without the snap-in effects, can be provided by means of force-controlled feedback system.[8] This technique may possibly provide the basis of the EFM measurement.

REFERENCES

1. Mukasa, K., Hasegawa, H., Tazuke, Y., Sueoka, K., Sasaki, M. and Hayakawa, K. (1994). *Jpn. J. Appl. Phys.*, **33**, 2692.
2. Mukasa, K., Sueoka, K., Hasegawa, H., Tazuke, Y. and Hayakawa, K. (1995). *Material Science and Engineering B.*, **31**, 69.
3. Nakamura, K., Hasegawa, H., Oguchi, T., Sueoka, K., Hayakawa, K. and Mukasa, K. (1997). *Phys. Rev. B.*, **56**, 3218.
4. Ness, H. and Gautier, F. (1995). *Phys. Rev. B.*, **52**, 7352.
5. Wang, C.S. and Freeman, A.J. (1981). *Phys. Rev. B.*, **24**, 4364.
6. Ohnishi, S., Freeman, A.J. and Weinert, M. (1983). *Phys. Rev. B.*, **28**, 6741.
7. Martin, Y., Williams, C.C. and Wickramasinghe, H.K. (1987). *J. Appl. Phys.*, **61**, 4723.
8. Jarvis, S.P., Yamada, H., Yamamoto, S.-I., Tokumoto, H. and Pethica, J.B. (1996). *Nature*, **384**, 247.

17. OBSERVATION OF SURFACE MAGNETISM OF MAGNETIC MULTILAYERS

GUIDO EILERS*

Spin Investigation Team of CREST, JST, Graduate School of Engineering, Hokkaido University, Sapporo 060-8628, Japan

ABSTRACT

This contribution contains information on what kind of work is planned to search for new magnetic materials in the course of the CREST project of the spin investigation team at Hokkaido University. The work will be mainly concerned with the interdependence of layer morphology and magnetic properties of metallic multilayers and the investigation of surface magnetism of dilutd magnetic semiconductors.

Keywords: surface magnetism, metallic multilayers, diluted magnetic semi-conductors.

INTRODUCTION

Research on the microscopic properties of magnetic materials became important for the development of high density recording media. In the course of this work the surface spin state will be examined on atomic scale. This will provide new insight into the interaction of localized magnetic moments at the surface of magnetic materials providing a first step for the development of nano-scale magnetic materials.

In research on applications of magnetic recording materials the question, how interaction between magnetic moments is working on a microscopic scale is of great importance. Similarly is the problem of the interaction mechanism between band electrons and localized moments in diluted magnetic semiconductors (DMS) of interest. Accordingly, this work will concentrate on the following two topics.

* Tel.: +81-11-706-6539, Fax: +81-11-706-7803, E-mail: guido@nano.eng.hokudai.ac.jp

MAGNETIC PROPERTIES AND SURFACE ROUGHNESS IN MAGNETIC MULTILAYERS

Though the properties of Fe/Cu as well as Co/Cu Multilayers have been investigated to considerable detail, the magnetic behavior of these materials on a microscopic scale is not fully understood. In the examination of fcc-Fe/Cu artificial superlattices it was found that the magnetic moment per Fe atom depends on the roughness of the Cu-substrate surface. (Kida, 1996) Figure 1(a) and (b) show Reflection High Energy Diffraction (RHEED) images of multilayers [x ML fcc-Fe/10ML Cu(001)]$_{20}$ set A and B with 20 periods grown at a substrate temperature of 320 K on Cu(001) buffers.

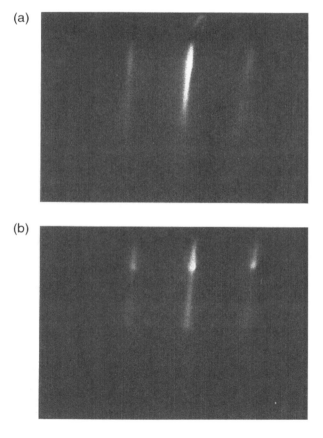

Figure 1 RHEED images of the 20th Fe layer, figure (a) and (b) show the image of sample set A and set B, respectively. For details see text.

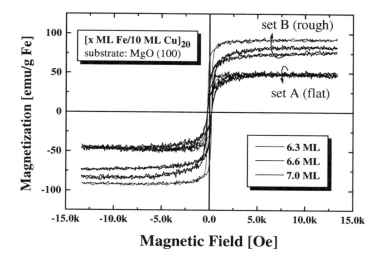

Figure 2 Magnetization of [x ML fcc-Fe/10ML Cu]$_{20}$ multilayers at room temperature, where x is the number Fe monolayers.

The 100 nm Cu buffers were prepared under the following conditions on MgO (001) substrates: the first 20 nm of the buffers of sample set A are grown at a substrate temperature of 670 K and the remaining 80 nm at 420 K, after that the Cu buffers were annealed for 30 min at 670 K; the entire 100 nm Cu buffers of set B were grown at a temperature of 670 K, no annealing was applied for set B. Since the streak length in RHEED diffraction images can be used as a measure of the surface roughness we conclude that the samples of set A have a much smoother surface than the samples of set B. The different surface condition of the two sets results in different magnetic properties as shown in Figure 2. It is clearly seen that the flatter surface leads to a smaller saturation magnetization at room temperature. The reason for this behavior will be studied on atomic scale with the following methods. The surface spin-state and topology of MBE grown fcc-Fe/Cu multilayers will be examined employing a spin polarized STM (SP-STM) with a GaAs-tip, gaining a spin resolved density of states with atomic scale resolution. From these results the occurrence of High- and Low-Spin state areas can be related to the topological structure.

Microscopic roughness also has strong impact on the properties of Co/Cu multilayers known to show a giant magnetoresistance (GMR) effect based on the antiferro-magnetic coupling between the magnetic Co layers and the non-magnetic Cu layers. In this research the microscopic origins of the GMR-effect will be examined experimentally through an observation of the surface spin-state with atomic spatial resolution. Since the coupling between the magnetic layers in Cu/Co multilayers was found to change periodically between antiferromagnetic and ferromagnetic, multilayers with a wedged non-magnetic (Cu) layer will be prepared to study the surface spin state at the border of domains of opposite spin polarization employing the SP-STM. An investigation of the relation of the spin-state with roughness and mixing at the layer interface through examinations of multilayer cross sections are planned.

A theoretical simulation of the experimental results will lead to a detailed understanding of the surface spin-state of magnetic multilayers and enable to control the state with a spatial resolution down to the atomic scale, supporting the development of nano-scale magnetic devices.

SURFACE SPIN STATE OF DILUTED MAGNETIC SEMICONDUCTORS

In DMS a fraction of cations is substituted at random by magnetic ions (Mn, Fe, Co). It is planned to study the interaction between the localized magnetic moments of these magnetic ions at the DMS surface and their influence on electrons in the vicinity. Experiments on the surface spin structure of DMS employing the SP-STM and the exchange force microscope (EFM) will achieve insight to the interaction between the magnetic ions at the surface of DMS. This will help to shed light into the problem of the atomic scale structure of bound magnetic polarons (BMP) in DMS. (Wolff, 1988) Another problem to be examined in this research is the scattering of conduction electrons at BMPs.

As a final goal the described research areas shall be combined to investigate the feasibility of hybrid materials of DMS and magnetic multilayers.

ACKNOWLEDGEMENTS

The help with preparations and magnetic property measurements of the multilayer samples by M. Matsui and Y. Yamada of the Department of Engineering at Nagoya University is gratefully acknowledged.

REFERENCES

Kida, A. and Matsui, M. (1996). Atomic Rearrangement and Magnetism of [Fe/Cu] Multilayers on Cu(001), *J. Phys. Soc. Jap.*, **65**(5), 1409.
Wolff, P.W. (1988). Bound Magnetic Polarons. In *Diluted Magnetic Semiconductors*, edited by Furdyna, J.K. and J. Kossut, Chap. 6, 413.

18. DESIGN OF MOLECULAR SPIN DEVICES: CONTROL OF SURFACE SPIN STATES

MAKOTO SAWAMURA* and KOICHI MUKASA

CREST/JST Spin Investigation Team,
Nanoelectronics Laboratory, Graduate School of Engineering,
Hokkaido University, Kita-13, Nishi-8, Kita-ku, Sapporo 060-8628, Japan

ABSTRACT

Advent of scanning tunneling microscope (STM) enabled us to observe and manipulate surface atoms by using an STM tip. Some theories for atom observation and manipulation under the STM environment were proposed. However, due to the restriction of calculational methods, spin properties of the whole system including surface atoms, a sample substrates and an STM tip are ignored. We investigate surface electronic states using molecular orbital theory which involves spin states explicitly. Based on the study on the behavior of surface electronic states, we found spin multiplicity of atoms plays an important role to determine the potential energy surfaces of a surface absorbed atoms between an STM tip and a sample. Taking advantage of these peculiar properties, we propose a mechanism for a spin device.

INTRODUCTION

Recently, spin electronic states have been found to play an important role under an STM environment.[1] Especially, spin-polarized tunneling has been reported to be observed.[2,3,4] Circularly polarized laser beamed on a zinc-blend crystal such as GaAs excites electrons to spin-polarized states. This mechanism was applied to a Spin-Polarized (SP) STM. Circularly polarized laser beam excites the electrons in an STM tip, and the excited spin-polarized electrons are transferred from the tip to the sample surface when the sample is positively biased. However, the unoccupied molecular orbitals of the sample surface can receive electrons with allowed spin states according to the selection rule. Taking advantage of such a characteristics, we may be able to observe surface spin states together with electronic states with STM.

* Tel.: 81-11-706-6539, Fax: 81-11-706-7803, E-mail: sawamura@nano.eng.hokudai.ac.jp

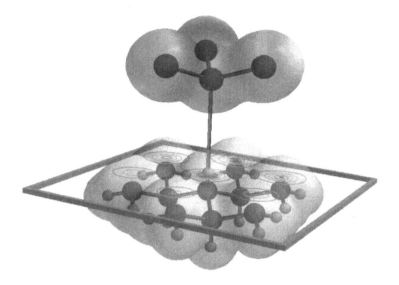

Figure 1 Cluster model for an STM. The tip is comprised of four Al atom, the surface of ten Si atoms and the absorbed atom is hydrogen. Dangling bonds of Si atoms are terminated by hydrogens. Transparent clouds are equi.

However, there is still some difficulty: new molecular orbitals are created between an STM tip and a sample. Therefore, the spin electronic states must be clearly derived and finally deconvoluted when we reveal the surface spin-electronic states. In this short article, we will show spin dependent surface properties through potential energy surfaces of an absorbed hydrogen atom on a silicon surface under an aluminum tip and a nanostructure on a surface sensitive to spin polarization will be presented to behave as a spin device.

SPIN-MULTIPLICITY-DEPENDENT POTENTIAL ENERGY SURFACES FOR AN ABSORBED ATOM UNDER AN STM

Advent of STM has enabled us to observe surfaces with atomic resolution. Then various applications were proposed and atom manipulation may be one of the most exciting possibilities.[2,5] It not only provoked researcher's interest, but it revealed fundamental problems in atomic processes of desorption or absorption on surfaces under strong electric field[1,6] and/or under an STM environment.[6] Such processes are clearly shown by potential energy surfaces for the absorbed atom.

We studied potential energy surfaces for a hydrogen atom absorbed on a silicon surface under an aluminum STM tip. We found potential energy surfaces change drastically according to the spin multiplicity of the whole system.

We constructed a cluster model for a tip, a sample and an absorbed atom as a whole (Figure 1). The tip is in a pyramidal structure, comprised of four aluminum atoms with the axis parallel to (111). The sample is a silicon surface (111) with four silicon atoms with the dangling bonds terminated by hydrogens. An absorbed hydrogen atom is located on the top of the silicon atom at the center. Figure 1 indicates ten silicon atoms, however, the following calculations were performed for the four silicon atoms in the surface. The bulk atomic distances were used.

Cluster models can reveal the spin states explicitly, however, they cannot describe characteristics due to delocalized electrons. But, in cases of desorption and absorption, electric current plays only secondary role[2] and the potential energy surfaces are determined by localized electrons.[1]

Calculations to determine the potential energy surfaces are done in such a way that the total energy of the whole system is calculated at each coordinate, as an absorbed atom moves from the surface to the tip. We show the potential energy surfaces for different spin multiplicity, singlet, triplet and quintet.

Unrestricted Hartree-Fock calculations with relativistic effective core potentials were employed. Gaussian94, Gaussian92 and Spartan ver. 4 were used on Silicon Graphics O2 and IBM Aptiva.

Figure 2 shows three potential energy surfaces. Each curve was obtained for different spin multiplicities and shows two potential wells on the both

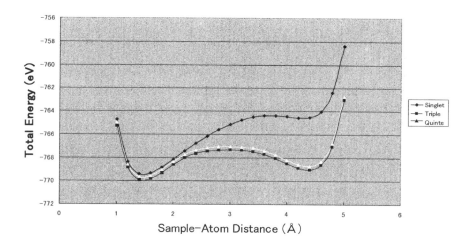

Figure 2   Potential energy surface for a hydrogen absorbed atom on a silicon surface below an aluminum STM tip.

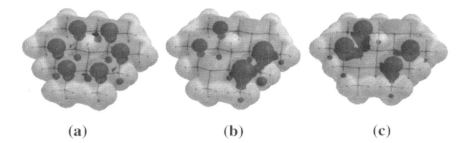

(a)   (b)   (c)

Figure 3 Cluster model of a silicon surface with an absorbed hydrogen. Solid clouds indicate HOMOs. Low spin states have higher Energy. (a) Total energy for singlet $(TE_{singlet})$ is 5442.377355 Hartree, (b) $TE_{triplet}$ −5442.529572 Hartree and (c) $TE_{quintet}$ −5442.554133 Hartree.

sides. The tip-sample distance is 7 Å. In case of singlet, the potential energy surface is inclined to the silicon side. However, in cases of triplet and quintet, the potential energy surfaces are nearly degenerated and show two potential wells clearly. These graphs indicate that if the numbers of alpha and beta spins are equal (singlet), the potential well of the silicon sample side is deeper than the well of the aluminum tip. Therefore, the absorbed atoms mostly stay on the silicon surface. On the other hand, the absorbed atom could stay on each side when the spin-multiplicity is triplet or quintet. The barrier height is 2–3 eV. In the latter case, the numbers of alpha and beta spins are not equal any more. The spin-multiplicity of the STM system drastically changes the potential energy surfaces for the absorbed atom.

We suggest if tunneling current is spin-polarized, the potential energy surface for the absorbed atom will change largely.

A PROPOSAL FOR A SPIN DEVICE

The previous section suggests that when the symmetry of the spin statistics of the system breaks down, surface characteristics may be changed. If tunneling current from an STM tip is polarized, electrons with a certain spin state may be allowed. We construct a simple nanostructure on a silicon surface in simulation. Figure 3 is a cluster model of a silicon surface (111) comprised of 19 silicons with a hydrogen atom absorbed on the top of it. The dangling bonds of the silicon atoms are terminated by hydrogens. We applied the above calculations to this structure. We found that as the spin multiplicity is higher, the total energy is lowered. The system is stabilized with a higher spin state. However, Figure 4 shows completely opposite

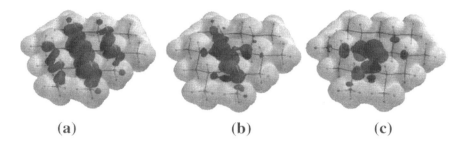

(a)   (b)   (c)

Figure 4 Cluster model of a silicon surface with six absorbed hydrogens. Solid clouds indicate HOMOs. High spin states have higher energy. (a) Total energy for doublet ($TE_{doublet}$) is -5445.966860 Hartree, (b) $TE_{quartet}$ -5445.659071 Hartree and (c) TE_{sextet} -5445.373382 Hartree.

properties. The cluster model in Figure 4 consists of the same silicon surface with six hydrogen absorbed atoms located hexagonally on the silicons. The cluster model shows that as the spin multiplicity is higher, the total energy is hightened.

We see the surface nanostructure, which is sensitive to spin electronic states, determines the properties of the surface. In cases of the cluster model in Figure 3, the spin of the system is polarized. However, the cluster in Figure 4 has nearly equal numbers of alpha and beta spins at the lowest energy electronic state. We may be able to design a surface nanostructure to produce spin characteristics.

CONCLUSIONS

We found spin-multiplicity plays an important role to determine surface desorption curves under an STM environment. Spin polarized states can be created by carefully constructing a surface nanostructure. It may behave as a spin device. However, we do not know yet the guiding principles to design spin devices using peculiar molecular orbital properties nor to predict spin characteristics without calculations. The spin-multiplicity dependent surface characteristics shown under an STM environment suggests that investigation of spin-electronic states of a system will indicate the spin-multiplicity and accordingly, in the future, will lead to nanostructures with spin-polarized states.[7]

REFERENCES

1. Sawamura, M., Tsukada M. and Aono, M. (1993). *Jpn. J. Appl. Phys.*, Vol. **32**, 3257.
2. Wiesendanger, R. (1994). *Scanning Probe Microscopy and Spectroscopy* (Cambridge University Press).
3. Sueoka, K. *et al.* (1993). *Jpn. J. Appl. Phys.*, **32**, 2989.
4. Alvarado, S.F. *et al.* (1992). *Phys. Rev. Lett.*, **68**, 1387.
5. Eigler, D. *et al.* (1991). *Phys. Rev. Lett.*, **66**, 1189.
6. Lang, N.D. *et al.* (1992). *Phys. Rev. B.*, **45**, 13599.
7. Sawamura, M., Maruyama, T. and Mukasa, K. (1997). *Proceedings of 2nd CREST Conference*, 4.

19. ELECTRON TUNNELING INTO DISORDERED MATERIALS

EIJI HATTA*

*Graduate School of Engineering, Hokkaido University,
Sapporo 060-8628, Japan*

The metal-insulator transition has been a subject of study for the last two decades and much progress has been made in understanding the physics in disordered materials. In a disordered material the electron motion is diffusive, hence, the electron screening is retarded and the Coulomb interactions among electrons are enhanced. The electronic properties of weakly disordered metals at low temperatures are governed by the effect of electron-electron interaction. This leads to a correction of the one-particle density of states (DOS). The correction was confirmed experimentally by tunneling experiments. Tunneling experiments showed that the DOS vanished at the metal-insulator transition. Above mentioned tunneling anomalies in the DOS have been widely observed in amorphous (magnetic) alloys, granular metals, and weakly disordered metals.

But, for example granular metals include essentially a correlated system of rather regularly alternating metallic and insulating regions. More attention has been paid to much more complicated and interesting structures, which would occur when the random solid mixture appears as a result of aggregation process or of a phase transition in some parts of the sample.[1]

On the other hand, it is of quite interest how the one-particle DOS of the disordered material can be affected by an additional disorder such as magnetic disorder. Some efforts have been made for studying amorphous magnetic materials using tunneling techniques; however, no direct evidence for the inherent effect of magnetic disorder (and not structural disorder) for the suppression of the tunneling conductance near zero bias voltage was obtained. Recently, the first direct evidence that the strong suppression on the tunneling conductance is largely due to the magnetic disorder is observed in $Ni_{1-x}Mn_x$ junctions.[2]

* Tel.: +81-11-706-6539, Fax: +81-11-706-7803, E-mail: hatta@nano.eng.hokudai.ac.jp

REFERENCES

1. Hatta, E. and Mukasa, K. (1997). *Solid State Commun.*, **103**, 235.
2. Hatta, E. and Mukasa, K. (1997). *Solid State Commun.*, **102**, 437.

INDEX